Palgrave Studies in Media and Environmental Communication

Series Editors

Drawing on both leading and emerging scholars of environmental communication, the Palgrave Studies in Media and Environmental Communication Series features books on the key roles of media and communication processes in relation to a broad range of global as well as national/local environmental issues, crises and disasters. Characteristic of the cross-disciplinary nature of environmental communication, the books showcase a broad variety of theories, methods and perspectives for the study of media and communication processes regarding the environment. Common to these is the endeavour to describe, analyse, understand and explain the centrality of media and communication processes to public and political action on the environment.

Daniel Binns · Rebecca Najdowski
Editors

Confronting the Climate Crisis

Activism, Technology and Ecoaesthetics

palgrave
macmillan

Editors
Daniel Binns
School of Media and Communication
RMIT University
Melbourne, VIC, Australia

Rebecca Najdowski
School of Art
RMIT University
Melbourne, VIC, Australia

ISSN 2634-6451 ISSN 2634-646X (electronic)
Palgrave Studies in Media and Environmental Communication
ISBN 978-3-031-89605-7 ISBN 978-3-031-89606-4 (eBook)
https://doi.org/10.1007/978-3-031-89606-4

This Palgrave Macmillan imprint is published by the registered company Springer Nature Switzerland AG
The registered company address is: Gewerbestrasse 11, 6330 Cham, Switzerland

For Adrian Miles
late and much-missed colleague, teacher, and friend

Acknowledgements

Chapter 6, 'A Close Reading of Climate-related Art: Aesthetics and Creative Engagement with the Structural Causes of the Climate Crisis', is an alternate version of part of Ulrike Hahn's PhD thesis, titled 'Imag(in)ing the Climate Crisis: The visual arts to the rescue?', at the Erasmus School of History, Culture and Communication at Erasmus University Rotterdam.

Chapter 8, 'Losing Our Heads: Finding Bodies in the Digital Anthropocene through Contemporary Painting Practice', is an alternate version of part of Rachel Magdeburg's PhD thesis, at the University of Wolverhampton.

Versions of Chapter 11, 'An Ecoaesthetic of Vegetal Surfaces: On Seed, Image, Ground as Soft Montage', have appeared in *Journal of Visual Art Practice*, 20(3), 2021, and in Abelardo Gil-Fournier and Jussi Parikka's book *Living Surfaces: Images, Plants, and Evironments of Media* (University of Minnesota Press, 2024). The work is reproduced here under Creative Commons Attribution 4.0 Licence for the journal article, and with the permission of the authors and publisher for the book chapter.

Contents

Notes on Contributors

Daniel Binns is Senior Lecturer in Media at RMIT University, Australia. He is a tinkerer-theorist exploring technology's impact on storytelling and media cultures, and a leading scholar on digital creativity and media transformation. Daniel has published on Netflix documentary style, drones, game engines and the evolution of media genres including the war film and superhero media. He is the author of *The Hollywood War Film* (2017) and *Material Media-Making in the Digital Age* (2021). In addition to his academic endeavours, Daniel has directed and produced network TV and his films have featured in over 20 festivals and streaming platforms.

Jacob Boivin is Lecturer and Doctoral Candidate in Communications at the École des Médias of the Université du Québec à Montréal (UQAM). His doctoral research project focuses on assessing the environmental impact of artificial intelligence (AI) from a social ecology perspective. He is a member of the Groupe de recherche sur l'information et la surveillance au quotidien (GRISQ) at UQAM. He is also associated with the Laboratoire d'éthique du numérique et de l'IA (LEN.IA). His research focuses more broadly on the sociological, ethical and philosophical analysis of AI, as well as on the political economy and political ecology of digital technologies.

Adrianus Both has a background in aquaculture research, specifically bivalve aquaculture. He has a MSc. (Aquaculture) from Memorial University of Newfoundland and a PhD (Marine Biology) from the University of Maine. His work has focused on mussel diet composition, how mussels interact with their local environment, and how their environment can affect their biochemical profile.

Hannah Brasier is a practice-based researcher interested in how noticing can be used to engage with the world ecologically. Her films combine the everyday, travel, landscape and environment to create ecologically conscious media. Hannah has multiple articles published in Studies in Documentary Film journal and her short film, Surface Levels (2023), was commissioned by the Mobile Innovation Network Association (MINA) in collaboration with Australian Environments on Screen.

Francesco Buscemi teaches media studies at the Catholic University of Milan. He pursued a PhD at Queen Margaret University, Edinburgh. Francesco taught at the universities of Stirling, Bournemouth, Venice and Como and won the *Santander Research Grant Fund*. He has been a TV writer, storyteller and journalist since the 1990s, wrote three academic books on food, culture and politics, and has recently published a novel about the imprisonment of conscientious objectors to military service in 1960s Italy.

Seán Cubitt is Professor of Screen Studies at the University of Melbourne. His publications include *The Cinema Effect, EcoMedia, The Practice of Light, Finite Media: Environmental Implications of Digital Technologies, Anecdotal Evidence: Ecocritique from Hollywood to the Mass Image* and *Truth: aesthetic politics 1.* Co-editor of *Ecocinema: Theory and Practice 2* and series editor for Leonardo Books at MIT Press, the history and philosophy of media, ecopolitical aesthetics and media arts and technologies.

Stephanie Farnsworth is a lecturer in Media and Communications at the University of Sunderland. Her research specialises in mutant, monster and body horror. She is the founder and co-Chair of the academic network MultiPlay which takes a multidisciplinary approach to video games studies, and is Chair of MultiPlay's editorial board. She is editor of the upcoming Bloomsbury collection *games That Haunt Us*.

Adalberto Fernandes is a researcher at the Research Centre for Tourism, Sustainability and Well-Being (CinTurs) at the Universidade do Algarve, Portugal. He has authored journal articles, chapters and presentations on communication, media, journalism, participation, festivals, events, trust, risk, science and health communication, political communication, censorship, biopolitics, bioethics, memes, cinema, environment, multiverse, critical discourse analysis, visual analysis and Michel Foucault.

Abelardo Gil-Fournier is an artist and researcher. With a Bachelor in Physics, he holds a PhD in Art from the Winchester School of Art (UK). He has worked as a researcher at FAMU in Prague (Czech Republic) and is currently a Leonardo grantee from the BBVA Foundation. His work has been exhibited internationally. He is the author, together with Jussi Parikka, of the book *Living Surfaces. Images, Plants and Environments of Media* (MIT Press 2024).

Ulrike Hahn is a PhD candidate at Erasmus University Rotterdam. Her PhD focuses on climate-related visual art. She is also working on creative projects (photography, short films, paintings) and has given several artistic workshops and presentations on art, circularity, climate change and sustainability. She won the NWO Synergy Award for her project on research-based comics and gives participatory workshops guiding researchers through the creation of their own research comics.

Charlie Hewison holds a PhD in History and Semiology of the Image from Université Paris Cité, and has lectured in cinema and screen cultures at institutions including Université Paris Nanterre, Université Paris 3 Sorbonne Nouvelle, and Université Paris-Cité. Charlie specialises in ecocritical approaches to audiovisual media, materialist film theory, and experimental cinema. He co-edited *Écocritiques: Cinéma, audiovisuel, arts* (2023) and *Cinématérialismes: Nouvelles approches matérialistes de l'audiovisuel* (2024). Charlie serves as Artistic Director of the Festival des cinémas différents et expérimentaux de Paris and is a member of the selection committee for experimental cinema distributor Light Cone.

Rachel Magdeburg is a visual artist and writer based in the UK. She is a Lecturer in Fine Art (Painting) and completed her PhD at the University of Wolverhampton in 2023 on painting and the concept of the Anthropocene. Rachel has held exhibitions and residencies across the UK and abroad, and has published both creative and academic writing including the article 'Self-Capture in the Anthropocene: The Expedition Paintings of Alexis Rockman' for *View. Theories and Practices of Visual Culture* (2018).

Wyatt Moss-Wellington is Senior Lecturer in Digital Storytelling at the University of New England. He is the author of *Cognitive Film and Media Ethics* (Oxford UP, 2021) and *Narrative Humanism: Kindness and Complexity in Fiction and Film* (Edinburgh UP, 2019).

Rebecca Najdowski is an artist-researcher at RMIT University. Through experimental approaches to photographic media, Rebecca investigates the mediation of ecological systems and environmental transformation. Her creative research spans analog experimentation, 3D photogrammetry, and generative AI to develop novel frameworks that question conventional representations of nature. Her work has been presented at the Museum Belvédère (The Netherlands), the Museum of Australian Photography, and Athens Digital Art Festival (Greece).

Daniel Palmer is a writer and Professor in the School of Art at RMIT University. His book publications include *Photography and Collaboration: From Conceptual Art to Crowdsourcing* (Bloomsbury, 2017); *Digital Light* (Open Humanities Press, 2015), edited with Sean Cubitt and Nathaniel Tkacz; and *The Culture of Photography in Public Space* (Intellect, 2015), edited with Anne Marsh and Melissa Miles.

Lili Pâquet is Senior Lecturer in Writing at the University of New England, with research interests in rhetoric, publishing studies and crime writing. Her books are *Crime Fiction from a Professional Eye* (2018) and the co-edited *True Crime and Women* (2024), and her poetry appears in *Cordite Review*, *Antipodes*, *Australian Poetry* and other publications.

Jussi Parikka is Professor in Digital Aesthetics and Culture at Aarhus University in Denmark. Parikka also holds a (visiting) professorship at University of Southampton (Winchester School of Art). In 2021 he was elected as member of Academia Europaea. In addition to several earlier books, he is the co-author of *The Lab Book: Situated Practices in Media Studies* (2022). Other recent books include *Operational Images* (2023) and the co-authored *Living Surfaces* (with Abelardo Gil-Fournier, 2024).

Ana Peraica is a theorist and photographer whose work focuses on post-digital photography, including generative and automated image forms. She coined the concept of photogenetics to explore the photographic recording of species extinction. Author of The Age of Total Images and Culture of the Selfie, she publishes widely across journals and anthologies.

Based in Split, Croatia, she also runs her family's historic photo studio and advocates for living heritage within UNESCO World Heritage sites.

Patrick Prax is Associate Professor in Game Design at Uppsala University. His academic and design work centers around systemic approaches to climate change and critical game and media literacy education. He has spoken at GDC about teaching sustainability in game education and is editor of a forthcoming book about making the games industry more sustainable. His games examine the connection of extreme wealth and climate obstruction and search for a path to material change.

Sieglind Wallner Hahn works in natural resources and sustainable development at the Department of Earth Sciences, Uppsala University, Sweden. Her research aims to advance social-ecological sustainability in coastal resource use. She is currently working on interview studies with small-scale fishers in Sweden and East Africa that highlight how human-induced changes to the marine ecosystem are impacting their livelihoods, and ways forward that can support their way of living in the long term.

List of Figures

CHAPTER 1

Navigating Catastrophe: Mapping Climate Crisis Responses Through Technology, Values, and Creative Practice

Daniel Binns and Rebecca Najdowski

Ebbs and Flows: Context, Relations, Values

In February 2024, the international Subcommission on Quaternary Stratigraphy voted against a proposal to codify the present epoch as the Anthropocene (Witze 2024). This decision appeared to bring to a close some fifteen years of debate about the measurement of humanity's impact on the geological record.[1] Members of the Subcommission were quick to point out, though, that this in no way diminished the reality of the planet's present situation. Interviewed for a New York Times article on the matter, environmental scientist Erle C. Ellis offered some reassurance:

D. Binns (✉)
School of Media and Communication, RMIT University, Melbourne, VIC, Australia
e-mail: daniel.binns@rmit.edu.au

R. Najdowski
School of Art, RMIT University, Melbourne, VIC, Australia

D. Binns and R. Najdowski (eds.), *Confronting the Climate Crisis*, Palgrave Studies in Media and Environmental Communication,
https://doi.org/10.1007/978-3-031-89606-4_1

> This was a narrow, technical matter for geologists, for the most part. This has nothing to do with the evidence that people are changing the planet. The evidence just keeps growing. (cited in Zhong 2024)

Further to this, members of the Subcommission also stated that the Anthropocene will still be categorized—in geological terms—as an 'event' rather than an 'epoch'. Epochs, like the Pliocene (5.333 million to 2.58 million years ago) or the Holocene (11,700 years ago to present), are geological time units defined by global events or transitions. The Miocene, for instance, is delimited by gradual cooling of the planet towards a number of ice ages (Steinthorsdottir et al. 2021). The Oligocene epoch continued this trend towards the global climate and geography that is more familiar today, being characterized by global growth of grasslands, and the diminishing of tropical forests to around the equator (Prothero 2005: 475–6). Geological events, by contrast, can range in duration from a second up to millions of years, sometimes spanning multiple epochs. These events can be physical or biological, such as a footprint, earthquake, or the Great Oxidation Event of 2.4 billion years ago (Gibbard et al. 2022: 3).

As Gibbard et al. put it:

> Defining the Anthropocene as a geological event allows it to be more closely connected with its diachronous stratigraphic evidence, which includes a wide range of deposits and paleoenvironmental proxies that attest to human transformations of environmental processes at local to global scales. (3)

While the planet is undoubtedly changing due to human impact, there is an ebb and flow to public consciousness and willingness to act. Huge numbers turned out in support of School Strike for Climate, an international youth-led activist movement demanding action from governments to curb environmental change and to transition to clean energy. In 2019, nearly 4000 strikes across 150 countries drew many hundreds of thousands of people into the streets (Haynes 2019). A global pandemic of the virus COVID-19 put into sharp relief the human footprint on the urban environment, when cities, roads, and waterways emptied during lockdowns. This absence allowed the non-human to return, allowing an unprecedented glimpse at how ecosystems adapt without constant human interference. Some species benefited: a herd of Kashmiri goats in Wales moved into the town of Llandudno for several days, where previously they might only shelter briefly during bad weather (Reuters 2020); a couple of striped dolphins took a turn through Venice's gondola-free canals (Machemer 2021); and in Albania, previously struggling populations of curly penguins and pink flamingos had a breeding boon in areas suddenly free of tourists (BBC 2020).

As COVID lockdowns continued into 2021, the United Nations convened in Glasgow for their Climate Change conference, known as COP26. The primary result of COP26 was an agreement by 197 participant nations to reduce reliance on fossil fuels generally, and coal specifically (COP26 2021). Russia's 2022 invasion of Ukraine caused an energy crisis that led some 2021 signatories to revert to coal and gas. By COP28 in 2023, many nations had neither altered their policies nor implemented their COP26 commitments. The financial and political means to make a smooth transition to renewables—or at least to reduce reliance on destructive forms of energy—remains difficult, particularly in regions struggling with other social, political, and economic factors. At COP28, participants declared a commitment to prevent climate change from increasing conflict, and to reduce "suffering created by the effects of climate change, environmental degradation and biodiversity loss" (COP28 2023). 130 countries also signed a pledge to prioritize renewables and triple renewable energy capacity by 2030 (ibid). Whether these pledges will be enacted remains unknown and contentious: every annual UN Climate Change Conference attracts large numbers of protesters that have their doubts.

Alongside the political, there are massive technological and industrial changes that have had huge ramifications for any form of conservation, safeguarding, or repair of the natural world. Blockchain and cryptocurrency had its first peak in 2021–22, with Bitcoin peaking in April 2021 at around US$63,500 (S. 2024). As use of blockchain grew, and the computational power required to 'mine' or generate blockchain tokens like Bitcoin also increased, energy expenditure skyrocketed. A September 2021 report estimated that the annual energy required to create Bitcoin alone was greater than that needed to power the country of Finland (Huang et al. 2021). Isolated efforts to produce cryptocurrency sustainably, like Solana and Ethereum, have had mixed results (Musharraf 2024; Ho 2023). There was an overall downturn of the cryptocurrency market following the FTX collapse in 2022, but the technology and the speculative market still exists, as do the attendant environmental impacts. In fact,

Bitcoin hit its all-time peak in November 2024, following the re-election of cryptocurrency supporter Donald Trump to the US Presidency (Lang et al. 2023; Aaron S. 2024; Schouten 2024).

After the crypto downturn, though, many of the servers and data centres previously used for mining were refurbished for driving generative artificial intelligence (genAI) systems (Taylor 2024; Canny 2023). The hype around genAI peaked in 2022–24, as companies like OpenAI, Anthropic, and Mistral released increasingly advanced large language models (LLMs) and media generators. Existing companies like Meta and Alphabet entered the industry, too, leveraging their established research teams and computational capacity to further the technology. As summarized by Butler and Lupton (2024: 15), it is the training of these genAI models that is the most resource- and energy-intensive part of their operation, with the preparation of OpenAI's GPT-4 estimated to have consumed approximately 300 tonnes of carbon (Kumar and Davenport 2023). But of course, other costs are incurred, with their own impacts, from materials sourcing and parts manufacture through to the huge amounts of cooling required to keep data centres running efficiently (Butler and Lupton 2024: 15). There are a few upsides, though, with some AI research hinting at promising developments including biodiversity preservation and nature conservation, as well as climate change science at scale (Butler and Lupton 2024: 5). Much work remains, though, to make these enormous resource-intensive systems in any way sustainable.

All of the foregoing political, social, and technological developments reflect the complex and varied nature of human relationships with the environment and other non-human beings. These relationships are shaped by the values people hold, be they intrinsic, instrumental, or relational. Intrinsic values denote the inherent worth of nature, valuing it for its own sake, independent of its utility to humans (Stålhammar and Thorén 2019: 1203). Instrumental values, on the other hand, consider nature primarily as a means to an end, i.e. in terms of its usefulness to humans as a resource, service, or other benefit (Arias-Arévalo et al. 2017: 43). Relational values, emerging as a significant third category, emphasize the importance of the relationships and responsibilities that people have towards nature, seeing these bonds as central to their identity and well-being. This concept challenges the conventional dichotomy between intrinsic and instrumental values by highlighting that human interactions with nature are often rooted in a complex web of emotional, cultural, and social connections (Stålhammar and Thorén 2019: 1202–1204). For instance, in Australian contexts, there is a significant relational value placed on the environment, where natural features are tied to national identity and cultural memory (Binns 2021: 118–19). Understanding these diverse values is obvious crucial in navigating and addressing environmental challenges, but also in approaching the situation from critical, creative, and philosophical positions.

The Anthropocene Elsewhere

The Anthropocene has been met with varying responses across disciplines, particularly in the humanities, arts, and media studies, which have embraced the term to reframe and expand the exploration of human and non-human interactions. This interdisciplinary engagement has sparked discussions that are fundamentally concerned with understanding and responding to the ecological crises we now face. The concept of the Anthropocene has inspired new methodologies, including new materialism and media materialism, to probe how art and media, stories, objects, and technologies are deeply entangled with ecological processes.

New materialism, per Karen Barad (2007, 2014) and Jane Bennett (2010), emphasizes the agentic capacities of matter, decentring the human in favour of recognizing how all entities, human and non-human, are enmeshed in broader ecological networks (Brice and Thorpe 2024: 359–60). This ontological shift (or ethico-onto-epistemological shift, to use Barad's term) challenges the logocentric and anthropocentric paradigms that have traditionally dominated qualitative research, offering instead a more inclusive understanding of materiality where objects and media are active participants in the co-creation of knowledge (Brice and Thorpe 2024: 361).

Within this framework, the term "Anthropocene" itself is reimagined not as a singular epoch marked solely by human impact but as a period defined by the entanglements of human and non-human actors. As scholars like Seàn Cubitt (2017), John Durham Peters (2015), and Jussi Parikka (2015) suggest, the media might be re-considered in relation to the Earth's geology (conceptually this time, rather than scientifically), with the extraction of rare earth materials, the disposal of electronic waste, or in terms of mediation or distribution as geological processes themselves. Parikka extends this notion by exploring how the extraction and accumulation of minerals for technological devices are processes that entangle human culture and temporalities with the deep time of the geological Earth (Parikka 2015: 4). Building on these ideas of environmental media, the concept of 'material witness' emerges as a way to understand non-human agency in the Anthropocene. Susan Schuppli introduced this concept to examine the evidentiary capacity of materials, "nonhuman entities and machinic ecologies" (2020: 3). Material witnesses possess non-human agency; within their materiality are traces of external ecological events and human interventions.

This rethinking of agency extends to creative practice, where artists are moving beyond anthropocentric views of nature and embracing broader ecological perspectives that acknowledge the agency of animals, plants, ecosystems, and geologic forces. Creative practice is a vital mode of inquiry and attunement to the Anthropocene because, as Heather Davis and Etienne Turpin argue, it is a 'sensorial phenomenon' shaped by visual imagery, models, and media (2015: 3–4). The climate crisis is largely mediated through images and media, both technical/operative and affective. Their ubiquity has contributed to the complex visuality of our ecological condition and the aesthetics of the Anthropocene (Mirzoeff 2014).

Expanding on a reconceptualization of agency and representation, Joanna Page advocates for 'planetary art' that transcends human agency and perception, attuning to the forces and temporalities of the planet itself, thereby enabling a more accurate understanding of human history within the broader context of planetary time (Page 2020). Scholars such as Nicholas Mirzoeff and T.J. Demos have similarly emphasized the importance of making long-developing environmental impacts legible and tangible (Mirzoeff 2014; Demos 2017). Artists are increasingly working within, and speculating about, the unsettling ecological and temporal conditions of 'slow violence' (Nixon 2011).

Much of this work highlights the entanglement of the colonial project with industrialization and extractive practices, which have contributed to current climate conditions and biodiversity loss. Scholars such as Demos, Page, and Kathryn Yusoff stress the necessity of engaging in decolonization to challenge Western-centric views of nature and the Anthropocene and respond to the climate catastrophe (Demos 2016; Page 2020; Yusoff 2018).

Finding Space for Collective Questioning

This book emerged out of several initiatives at RMIT University in Melbourne, Australia. The first of these was the Posthumanities Reading Group, begun in 2016 by interactive documentary theorist Adrian Miles. Following Adrian's tragic passing in 2018, the group reoriented around the eco_media conference series from 2019 to 2022. These events brought together scholars and practitioners at all career stages, with screenings and presentations of diverse research, media pieces, and works-in-progress. The project took on a life beyond RMIT in 2020 when, due to an entirely different global emergency, eco_media shifted online. This afforded the opportunity to hear new and diverse perspectives from interstate and abroad.

A culture of experimentation and divergent thinking emerged from these events, as well as the act of questing and questioning—"finding the grain of the world's becoming and following its course" (Ingold 2011: 92)—as a means of confronting global catastrophes of all kinds. This book is a reflection of these ideas, and the interdisciplinary collective that emerged around them, drawing from fields such as environmental humanities, media studies, creative practice research, and philosophy. The following chapters offer diverse methods for confronting to the climate catastrophe. Some demonstrate the vital role of activism and social movements in raising awareness and challenging systemic issues; others explore the potential of artistic and creative practices to ask difficult questions, evoke emotional resonance, and imagine alternative futures. What connects these approaches is an examination of digital media and emerging technologies—like podcasts, video games, machine learning, and extended reality (XR)—human ingenuity and technological innovation, for their potential to provide new perspectives, aid in creative interpretations, and facilitate collective action in addressing climate challenges.

One might approach these contributions through the lens of Celia Lury's concept of a 'problem space' (Lury 2021: 6). As described by Jussi Parikka and Abelardo Gil-Fournier in their chapter, a problem space is "where problems, methods, and concepts are dynamically re-examined and rearticulated and where the space of their composition is not a stable container where a thing is addressed but part of their making" (this volume). This collection itself can be understood as a problem space, blurring the boundaries between the problem being confronted and the very processes of grappling with it. Taken together, the chapters trace the diverse contours of the climate 'problem' through varied engagements, methodologies, and conceptual framings. The problem space enacted by this volume resonates with Donna Haraway's call to "stay with the trouble" of our current environmental predicament (Haraway 2016). Haraway argues against dismissing the challenges of the climate crisis with a strong faith in technofixes as well as the fatalistic position that it is too late to do anything meaningful about our predicament. Instead, she advocates for staying with the trouble of living within an ecological crisis. Climate change emerges not as an isolated event to be solved, but as an ongoing trouble, to be 'troubled' or problematized, in which our responsibility is through sustained engagement.

Overview of Contributions

Protest and Progress: Activism, Education & Ecopolitics

The opening section 'Protest and Progress: Activism, Education & Ecopolitics' offers four approaches to environmental education and communication in the face of the climate crisis. Across these approaches, there is a tension observed between individual and systemic perspectives on addressing climate change.

Focusing on an education environment, Patrick Prax, Sieglind Wallner Hanh, and Adrianus Both describe a university course module aimed at teaching students to think systemically about sustainability rather than focusing solely on individual responsibility. The course combines critical social science perspectives with natural science to contextualize environmental issues politically. While students showed progress in critical thinking, they still defaulted to frames of individual responsibility, highlighting the challenge of overcoming deeply ingrained cultural narratives.

Focusing on the utility of the body as a tool with and for activism, Francesco Buscemi considers corporeality and materiality in the strategies of young environmental protesters. Buscemi juxtaposes this embodied action with the immateriality of the 'metaverse' proposed by Meta and its contemporaries. For Meta CEO Mark Zuckerberg, the metaverse promises a new era of political cooperation; for protesters like the Fridays for Future movement, Meta is just one more contributor to catastrophe.

Adalberto Fernandes analyses memes critiquing the COP26 climate conference, arguing that memes represent a form of environmental communication that is part of the environment itself rather than separate from it. He suggests memes use a "hypercritical" strategy, employing fiction and humour to critique the real hypocrisy of climate politics. However, this approach is itself hypocritical as it relies on polluting media industries to spread its message.

Wyatt Moss-Wellington and Lili Pâquet analyse environmental podcasts, examining how they use emotional narratives and humour to inspire care for the environment and motivate listeners to take action. The case studies discussed—'Outrage + Optimism' and 'Sustainababble' utilize anger, hope, and humour to build a sense of community among listeners. This chapter's analytic method reveals how these audio works personalize activist rhetoric, empowering individuals to pursue climate action in ways that fit their own circumstances.

These opening chapters all tackle the challenge of shifting public discourse and understanding from individualized to systemic approaches to climate change. Memes, podcasts, and educational initiatives all grapple with how to effectively communicate the need for large-scale societal and economic changes while still empowering individuals to take meaningful action. Key to this communication and empowerment is emotion, which is highlighted across these chapters, from the use of humour in memes to the careful balancing of outrage and optimism in podcasts. Purely negative or anxiety-inducing messaging can be paralyzing, while overly positive framing may not convey sufficient urgency. This section underscores the ongoing struggle to reframe climate discourse away from individual consumer choices and towards collective, systemic action, while still maintaining hope and a sense of personal agency in the face of a global crisis.

Ecoaesthetic Practice & Analysis

The section 'Ecoaesthetic Practice & Analysis' examines how visual artists, photographers, and filmmakers are engaging with climate change and environmental issues through their creative work. It explores both the aesthetic strategies and the ways these images and artworks make visible unseen impacts, evoke emotional resonance, and confront with the complex political, social, and ecological dimensions of the climate crisis.

Ulrike Hahn provides an in-depth analysis of several climate-related artworks, examining how they employ aesthetics to engage with the political and structural causes of climate change. Hahn's analysis of works ranging from video essays and paintings to large-scale public installations emphasizes the importance of engaging with the complex social, political, and economic dimensions of climate change. The chapter underscores the need for transparency and research on the environmental impact of art practices and advocates for an aesthetics of sustainability that fosters sensitivity to complexity and interconnected patterns.

Unintentional climate documents emerge in Ana Peraica's analysis of ubiquitous photographs. By examining stock photographs, family snapshots and other everyday visuals, Peraica reveals how vernacular imagery can serve as a record of environmental changes over time. Peraica's concept of the Photographocene—a photographic timeline of climate change—frames her discussion of how everyday images mediate our perception of ecological collapse. She highlights how many photographs, particularly in professional and commercial contexts, aestheticize or omit environmental damage, delaying critical awareness. This chapter thus demonstrates the value of looking beyond traditional forms of climate documentation and highlights the potential for critical examination of vernacular imagery to uncover new insights about our changing planet.

Rachel Magdeburg's chapter examines how decapitation in painting symbolizes the mind-body disconnect in Cartesian dualism. She presents emoji as modern "severed heads," acting as proxies for human figures in digital interfaces. Magdeburg's work proposes that emoji brought into painting through the material presence and the indexical traces of artistic labour can allude to the physical, psychological, and existential ruptures of the Anthropocene.

Charlie Hewison examines experimental films that visualize normally invisible environmental threats like radiation and electromagnetic fields through material manipulation of the film itself. The chapter critiques the reliance on hypervisible spectacles in mainstream and eco-political cinema, arguing that such portrayals often fail to capture the insidious nature of gradual environmental decline. Instead, Hewison advocates for a "process cinema" approach, prioritizing material interactions and transformations over conventional representation, emphasizing the importance of representing less visible but pervasive threats.

Experimental documentary approaches that transcend spectacular disaster imagery are the focus of Hannah Brasier's chapter. She advocates for an approach that emphasizes "attuned noticing" and interactive filmmaking to move beyond traditional linear storytelling. Key methods include using arbitrary constraints and the camera's "indifference" for unbiased footage, using interactive software like Korsakow for non-linear video associations, and encouraging contemplative viewing through looping clips and user interaction. The chapter suggests that reducing authorial control and fostering deeper engagement through interactivity can create richer, multi-layered representations of environmental issues, promoting documentary practices that better reflect the complexity and uncertainty of an ongoing climate disaster.

Together, these analyses demonstrate the diverse and evolving ways image makers and visual artists are confronting the representational and conceptual challenges posed by climate change. They underscore art's capacity to offer new perspectives and modes of engagement with the climate crisis, moving beyond spectacular imagery to capture ongoing, distributed changes.

Emerging Technological Practices

The section 'Emerging Technological Practices' explore how emerging media technologies are being deployed to engage with environmental and ecological issues. These four chapters demonstrate how diverse forms of visualization and data processing can offer novel ways of representing and understanding environmental shifts.

Parikka and Gil-Fournier examine the intersection of plant growth, media surfaces, and environmental monitoring through their video project "Seed, Image, Ground." They consider how images of vegetal growth relate to technological surfaces like screens, exploring the temporalities and materialities involved in visualizing environmental processes.

Rebecca Najdowski and Daniel Palmer explore how emerging machine learning and AI technologies are being used by artists to generate speculative imagery related to climate change and the Anthropocene. They examine projects that use large datasets of photographs to produce new visualizations that move beyond conventional documentary approaches. Key examples include Refik Anadol's data-driven landscapes, Anna Ridler's handcrafted datasets, and Tega Brain, Julian Oliver, and Bengt Sjölén's critical AI simulations.

Using the video games *Mass Effect* and *The Last of Us* as case studies, Stephanie Farnsworth considers depictions of post-apocalyptic environmental collapse, looking at how bodies and landscapes are represented as "wasted" in these virtual worlds. The chapter considers how these games reflect anxieties about climate catastrophe while still operating within neoliberal frameworks.

Rewa Wright discusses how extended reality (XR) technologies are being used to create "bio-digital assemblages" that merge nature, technology and human agency in artistic practices addressing climate crisis. This allows for new conceptual and embodied engagements with environmental issues.

In sum, these chapters consider how technological mediation might enable comprehension of non-human temporalities and perspectives and prediction or visualization of climate futures; the works also consider how media technologies themselves are imbricated with ecological systems. Emerging media practices are enabling new modes of ecological perception, representation, and engagement, while also requiring critical examination of their own material and conceptual implications.

New Theoretical Horizons

The two final chapters pose philosophical questions that offer critical perspectives on the intersections of technology, ecology, and human subjectivity.

In a novel approach to ecological critique, Seán Cubitt's chapter embarks on three digressions using seemingly disparate motifs—mermaids, angels, and cats—as metaphorical lenses to unravel the complexities of disgust, terror, and the uncanny. Cubitt reflects on the nature of desire, the sublime, and the multifaceted relationships between humans, technology, and the environment, emphasizing the role of aesthetic experience and imagination in addressing ecological crises. Ultimately, Cubitt proposes an eco-aesthetic politics as a framework for confronting contemporary environmental challenges.

Building on Cubitt's notion of ecological and technological entanglements, Jacob Boivin proposes that technological mediation is the fundamental condition of our experience of the world. This chapter argues that understanding this condition, characterized by 'algorithmic mediation', is a necessary shift for ecological thinking and cultural change. The algorithmic condition, driven by the demand for data in the digital economy, creates a cyclical process that interconnects and amplifies geological, biological, and psychosocial impacts, costs, and destructions. Boivin contends that these systems profoundly influence our subjectivity but fail to fully capture the complexity of both our reality and the present ecological crisis.

Cubitt and Boivin argue for new ways of thinking about and engaging with the world in the face of ecological crisis, emphasizing the need for imaginative, interdisciplinary approaches that can apprehend the complexity of our current moment in anticipation of meaningful action.

*****

In a single volume, *Confronting the Climate Crisis* brings together research offering diverse insights on and approaches to the present and ongoing ecological catastrophe. Running through many chapters is an attending to the realities of digitally-mediated subjectivity, and how new technologies—when responsibly manufactured and deployed—can present new perspectives and afford creative interpretations of environmental issues.

This collection intentionally muddles the boundaries between technological potential, sense-making and storytelling, and critical awareness. This is an ensemble of works that stay with the trouble, speaking to each other, but also, we hope, opening dialogues with others working in this particular 'problem space'. As we move through successive climate conferences, technological shifts, and waves of public engagement, the need for such varied perspectives becomes increasingly clear.

Note

1. Several members of the group have, at the time of writing, expressed concern about the decision (Zhong 2024).

References

Arias-Arévalo, Paola, Martín-López, Berta and Gómez-Baggethun, Erik (2017), 'Exploring intrinsic, instrumental, and relational values for sustainable management of social-ecological systems', *Ecology and Society*, 22:4, https://doi.org/10.5751/ES-09812-220443.

Barad, Karen Michelle (2007), *Meeting the Universe Halfway: Quantum Physics and the Entanglement of Matter and Meaning*, Durham: Duke University Press, e-Duke books scholarly collection., https://doi.org/10.1515/9780822388128.

Barad, Karen (2014), 'Diffracting Diffraction: Cutting Together-Apart', *Parallax*, 20:3, https://doi.org/10.1080/13534645.2014.927623.
BBC (2020), 'Coronavirus: Wild animals enjoy freedom of a quieter world', BBC News, 29 April, https://www.bbc.com/news/world-52459487. Accessed 25 July 2024.
Bennett, Jane (2010), *Vibrant Matter: A Political Ecology of Things*, Durham: Duke University Press.
Binns, Daniel (2021), 'Attuning to the environment through media: Escape and incorporation through fire, plague and video game development software', *Journal of Environmental Media*, 2:1, https://doi.org/10.1386/jem_00043_1.
Brice, Julie E. and Thorpe, Holly (2024), 'New Materialisms, Material Methods, and the Research Process: A Creative Experiment in Cutting Together-Apart', *Leisure Sciences*, 46:3, https://doi.org/10.1080/01490400.2021.1981504.
Butler, Ella and Lupton, Deborah (2024), *Generative AI Technologies Applied to Ecosystems and the Environment: A Scoping Review*, Melbourne: ARC Centre of Excellence for Automated Decision-Making and Society, ADM+S Working Paper Series 2024, https://doi.org/10.60836/k7fv-7991. Accessed 10 September 2024.
Canny, Will (2023), *Crypto Miners Are Attempting to Diversify Into Other Business Areas: JPMorgan*, CoinDesk, https://www.coindesk.com/business/2023/08/21/crypto-miners-are-attempting-to-diversify-into-other-business-areas-jpmorgan/. Accessed 10 September 2024.
COP26 (2021), COP26 Keeps 1.5C Alive and Finalises Paris Agreement, UN Climate Change Conference UK 2021, https://ukcop26.org/cop26-keeps-1-5c-alive-and-finalises-paris-agreement/. Accessed 25 July 2024.
COP28 (2023), *COP28 Declaration On Climate, Relief, Recovery And Peace*, COP28: The UAE Consensus, https://www.cop28.com/en/cop28-declaration-on-climate-relief-recovery-and-peace. Accessed 22 November 2024.
Cubitt, Sean (2017), *Finite Media: Environmental Implications of Digital Technologies*, Durham: Duke University Press, A cultural politics book, https://doi.org/10.1515/9780822373476.
Davis, Heather and Turpin, Etienne (2015), *Art in the Anthropocene: Encounters Among Aesthetics, Politics, Environments and Epistemologies*, London: Open Humanities Press.
Demos, T.J. (2016), *Decolonizing Nature: Contemporary Art and the Politics of Ecology*, Berlin: Sternberg Press.
Demos, T.J. (2017), *Against the Anthropocene: Visual Culture and Environment Today*, Berlin: Sternberg Press.
Gibbard, P., Walker, M., Bauer, A., Edgeworth, M., Edwards, L., Ellis, E., Finney, S., Gill, J.L., Maslin, M., Merritts, D. and Ruddiman, W. (2022), The

Anthropocene as an Event, not an Epoch. *J. Quaternary Sci*, 37: 395–399. https://doi.org/10.1002/jqs.3416

Haraway, Donna (2016), *Staying with the Trouble: Making Kin in the Chthulucene*, Durham: Duke University Press.

Haynes, Suyin (2019), Students From 1,600 Cities Just Walked Out of School to Protest Climate Change. It Could Be Greta Thunberg's Biggest Strike Yet, *TIME*, https://time.com/5595365/global-climate-strikes-greta-thunberg/. Accessed 25 July 2024.

Ho, Charlyn (2023), 'One Year After The Merge: Sustainability Of Ethereum's Proof-Of-Stake Is Uncertain', *Forbes*, 11 October, https://www.forbes.com/sites/digital-assets/2023/10/11/one-year-after-the-merge-sustainability-of-ethereums-proof-of-stake-is-uncertain/. Accessed 10 September 2024.

Huang, Jon, O'Neill, Claire and Tabuchi, Hiroko (2021), 'Bitcoin Uses More Electricity Than Many Countries. How Is That Possible?', *The New York Times*, 3 September, https://www.nytimes.com/interactive/2021/09/03/climate/bitcoin-carbon-footprint-electricity.html. Accessed 10 September 2024.

Ingold, Tim (2011), 'The textility of making', *Cambridge Journal of Economics*, 34, https://doi.org/10.1093/cje/bep042.

Kumar, Ajay and Davenport, Tom (2023), *How to Make Generative AI Greener*, Harvard Business Review, https://hbr.org/2023/07/how-to-make-generative-ai-greener. Accessed 10 September 2024.

Lang, Hannah, Howcroft, Elizabeth and Wilson, Tom (2023), 'The crypto market bears the scars of FTX's collapse', *Reuters*, 3 November, https://www.reuters.com/technology/crypto-market-still-bears-scars-ftxs-collapse-2023-10-03/. Accessed 22 November 2024.

Lury, Celia (2021), *Problem Spaces. How and Why Methodology Matters*. Cambridge: Polity.

Machemer, Theresa (2021), Dolphins Spotted in Venice's Grand Canal—for Real This Time, *Smithsonian Magazine*, https://www.smithsonianmag.com/smart-news/dolphins-spotted-venices-grand-canal-real-time-180977403/. Accessed 25 July 2024.

Mirzoeff, Nicholas (2014), 'Visualizing the Anthropocene', *Public Culture*, 26:2, pp. 213–232.

Musharraf, Mohammad (2024), *Understanding the Environmental Impacts of NFTs and Greener Alternatives*, MoonPay, https://www.moonpay.com/learn/nft/nft-environmental-impact. Accessed 10 September 2024.

Nixon, Rob (2011), *Slow Violence and the Environmentalism of the Poor*, Cambridge, MA: Harvard University Press.

Page, Joanna (2020) 'Planetary art beyond the human: Rethinking agency in the Anthropocene,' *The Anthropocene Review*, 7:3, pp. 273–294.

Parikka, Jussi (2015), *A Geology of Media*, Minneapolis: University of Minnesota Press, Electronic Mediations.

Peters, John Durham (2015), *The Marvelous Clouds: Toward a Philosophy of Elemental Media*, Chicago, [Illinois] ; The University of Chicago Press, https://doi.org/10.7208/9780226253978.

Prothero, D R (2005), 'Oligocene' (eds R. C. Selley, L. R. M. Cocks and I. R. Plimer), in R. C. Selley, L. R. M. Cocks, and I. R. Plimer (eds), *Encyclopedia of Geology, Volumes 1 - 5*, Kidlington: Elsevier Ltd., pp. 472–8, https://app.knovel.com/hotlink/pdf/id:kt004ON8HG/encyclopedia-geology/oligocene-life.

Reuters (2020), 'Goats and coronavirus lockdown: Kashmiri goats invade deserted Welsh seaside resort of Llandudno', The Sydney Morning Herald, 1 April, https://www.smh.com.au/traveller/inspiration/goats-and-coronavirus-lockdown-goats-invade-deserted-welsh-seaside-resort-20200401-h1n1nk.html. Accessed 25 July 2024.

S., Aaron (2024), *Bitcoin Price History: What Was Bitcoin's Highest Price?*, BitDegree, https://www.bitdegree.org/crypto/tutorials/what-was-bitcoins-highest-price. Accessed 22 November 2024.

Schouten, Fredreka (2024), *The Crypto Industry Plowed Tens of Millions into the Election. Now, It's Looking for a Return on That Investment | CNN Politics*, CNN, https://www.cnn.com/2024/11/17/politics/crypto-industry-donald-trump-reelection/index.html. Accessed 22 November 2024.

Schuppli, Susan (2020). *Material Witness: Media, Forensics, Evidence*, Cambridge, MA: The MIT Press.

Stålhammar, Sanna and Thorén, Henrik (2019), 'Three perspectives on relational values of nature', *Sustainability Science*, 14:5, https://doi.org/10.1007/s11625-019-00718-4.

Steinthorsdottir, M., Coxall, H. K., De Boer, A. M., Huber, M., Barbolini, N., Bradshaw, C. D., Burls, N. J., Feakins, S. J., Gasson, E., Henderiks, J., Holbourn, A. E., Kiel, S., Kohn, M. J., Knorr, G., Kürschner, W. M., Lear, C. H., Liebrand, D., Lunt, D. J., Mörs, T., Pearson, P. N., Pound, M. J., Stoll, H. and Strömberg, C. A. E. (2021), 'The Miocene: The Future of the Past', *Paleoceanography and Paleoclimatology*, 36:4, https://doi.org/10.1029/2020PA004037.

Taylor, Josh (2024), 'Data gold rush: companies once focused on mining cryptocurrency pivot to generative AI', *The Guardian*, 26 January, https://www.theguardian.com/australia-news/2024/jan/27/tech-companies-shift-generative-ai-chatgpt. Accessed 10 September 2024.

Witze, Alexandra (2024), 'Geologists reject the Anthropocene as Earth's new epoch — after 15 years of debate', *Nature*, 627:8003, https://doi.org/10.1038/d41586-024-00675-8.

Yusoff, Kathryn (2018), *A Billion Black Anthropocenes or None,* Minneapolis: University of Minnesota.

Zhong, Raymond (2024), 'Are We in the "Anthropocene," the Human Age? Nope, Scientists Say.', The New York Times, 5 March, https://www.nytimes.com/2024/03/05/climate/anthropocene-epoch-vote-rejected.html. Accessed 15 May 2024.

PART I

Protest and Progress: Activism & Ecopolitics

CHAPTER 2

Teaching Sustainable Development: On Systemic Issues and Fake Excuses

Patrick Prax, Sieglind Wallner Hahn, and Adrianus Both

Introduction

This chapter discusses the case of a course module for an interdisciplinary university course that aims to teach a systemic perspective on sustainability questions. The course is based on the report "Making Peace with Nature" from the United Nations environment program (Environment, 2021). The report explains that the needed 'Transformation will involve a fundamental change in the technological, economic, and

P. Prax (✉)
Department of Game Design, Uppsala University, Visby, Sweden
e-mail: patrick.prax@speldesign.uu.se

S. Wallner Hahn
Department of Earth Sciences, Uppsala University-Campus Gotland, Visby, Sweden

A. Both
Kristineberg Marine Research Station, IVL Swedish Environmental Research Institute, Kristineberg, Sweden

D. Binns and R. Najdowski (eds.), *Confronting the Climate Crisis*, Palgrave Studies in Media and Environmental Communication,
https://doi.org/10.1007/978-3-031-89606-4_2

social organization of society, including world views, norms, values and governance' (Environment, 2021: 13). Using a systemic perspective, this chapter aims to emphasize that the needed transformation goes beyond individual consumer actions and questions of personal morality. From this perspective, climate change needs to be addressed through real, material changes in our global production systems, mechanisms of governance, and economies if there is to be any chance to avoid climate catastrophe. These changes need to be accompanied by cultural, societal, and normative changes, but those immaterial elements cannot stand alone.

The UN report also explains that 'some attractive and feasible actions can impede transformation' (p. 113). As a mechanism for this impediment the report explains that 'The changes that appear most feasible may be those that do not contribute to, or even impede, transformative change, for instance by retaining or even consolidating the power interests vested in the status quo' (p. 114). Here it says that 'Some individuals and organizations also have substantial stakes in maintaining the status quo. These vested interests may oppose changes that disrupt their livelihoods, market shares and future revenues' (p. 104). Stressing the responsibility of individuals exclusively and placing the responsibility for climate action at the feet of the individual customers on a marketplace exclusively, even in environmental education (Ferreira 2019), is here seen as creating the kind of impediment to the needed change mentioned by the UN report. Individualization disguises material and systemic problems as exclusively issues of individual virtue and character, and situates responsibility for climate change on individuals instead of the vested industrial interests that benefit from the status quo (Maniates 2001). It was further discussed if the individualization of climate responsibility then is not only an involuntary consequence of neoliberalism (Kent 2009; Maniates 2001), but a purposeful distraction and greenwashing by the fossil fuel industry (Delmas and Burbano 2011; Laufer 2003; Supran and Oreskes 2021). In a call for action, Maniates (2001: 44) states that 'the forces that systematically individualize responsibility for environmental degradation must be challenged.' In order to facilitate the required societal changes, it is necessary to re-frame the discussion away from individualization, especially when teaching sustainability (Ferreira 2019), and move towards an understanding of climate change as a symptom of global systemic dysfunction (Wals 2019). This chapter is a part of this effort of creating Education for Sustainable Development (ESD) that resists the frame of individual responsibility to instead focus on systemic perspectives.

The course module discussed how to distribute responsibility for environmental action as well as the role of consumer action in the required transformation. For example, the module discussed what responsibilities fossil fuel extracting companies, large-scale producers of goods, and the food industry have compared to individual consumers, and which impact could be expected from their transformations and actions.

Education for Sustainable Development (ESD)

ESD is recognized as an essential component of the Sustainable Development Goal (SDG) 4 on quality education and is a key enabler of all other SDGs. Through ESD, UNESCO strives to ensure that all generations understand the impact of climate change and are better equipped to act and protect resources, the environment, and the planet that sustains life (UNESCO 2019). This call for ESD is crucial, as humanity has become the dominant force of change on our planet, shaping the future of the Earth system as a whole, giving rise to a new epoch referred to as the Anthropocene. Our biosphere has been changed by human actions causing climate change, rapid loss of biodiversity, growing inequalities, and loss of resilience to deal with uncertainty and surprise (Folke et al. 2021; Biermann et al. 2012). People and nature are deeply intertwined, and the collective actions of humanity will affect whether future conditions are hostile or beneficial for human life and wellbeing. 'Whether humanity has the collective wisdom to navigate the Anthropocene to sustain a livable biosphere for people and civilizations, as well as for the rest of life with which we share the planet, is the most formidable challenge facing humanity' (Folke et al. 2021).

Strong interdisciplinary skills are required to understand and enable the required transformative change towards sustainability. As such, interdisciplinary skills and critical thinking must be fostered in younger generations. Higher education institutions and universities have a significant responsibility in providing education in the field of sustainability and can play a powerful role in creating and spreading crucial knowledge on how to tackle climate change, environmental degradation, over-consumption and social injustice and inequity. Holistic, interdisciplinary approaches are crucial to prepare future professionals with the scientific knowledge, skills, values, and motivations to take action for sustainability from their personal lives to the global scale.

Course Module Description

The aim of this section is to describe the course module in enough detail to allow replication. The module described in this section was part of a larger interdisciplinary course held at Uppsala University, Sweden, titled: "A Sustainable Future—Theory and Transdisciplinary Visions". The course aimed at providing students with an introduction to real-world complex sustainability challenges, while the goal of the module described in this section was to teach students to identify the discursive frame and foster critical thinking of proposed sustainability solutions and their capacity to enact necessary social-ecological transformations. The module discussed ethics, norms and values, political will, along with their unintended consequences that hinder sustainability transformations. Students were encouraged to consider the complexities of social, ecological, and economic systems with an emphasis balancing trade-offs when considering possible sustainability actions. This holistic system thinking approach was intended to provide students with tools to analyze and scrutinize offered initiatives to determine whether they were adequate to tackle sustainability challenges.

The interdisciplinary course module consisted of four parts: The first introduces students to the overarching question of whether the currently dominating discursive framing of climate change was around individual responsibility or systemic perspectives. It also introduced students to their course assignment, which asks students to explore what they consider meaningful climate mitigation strategies versus what they believe are distractions. This course section was taught by a social scientist and media researcher. The second and the third parts of the course offered students detailed examinations of specific sustainability challenges from social-ecological perspectives. Allowing them to explore sustainability issues and have conversations with experts about complexities and discuss possible solutions. These lectures were meant to provide a concrete counterpart to the more general and abstract conversation about the discursive frames. In this instance, these lectures were taught by researchers covering small-scale fisheries and aquaculture. One lecture presented a rich human picture, informed by ethnographic work, taking the students to the beach of East Africa with stories and images of small-scale fisheries. The other lecture navigated students through complex global aquaculture production data and used examples of food production methods, including their benefits, challenges, and trade-offs, to show where environmental, economic, and social sustainability issues intersected. The fourth and final course section consisted of presentations by students of possible productive approaches to climate change vs. deficient or superficial initiatives, followed by facilitated discussions.

Part One: Systemic Perspectives and Discursive Frames

This part of the course set up the conversation about discursive frames by discussing a case study where a polluting industry previously set the discursive frame. This case study introduced students to a well-studied example, plastic pollution, which benefits from hindsight and encourages them to think critically about the issue. Students were given reading material about the role plastic recycling plays as an excuse for plausible deniability by the plastic industry. They explored the case of the 'Crying Indian' advertisement, which frames plastic pollution as an individual rather than as a systemic issue (Andersen 2013; and as an accessible journalistic source Lerner 2019).

This critical lens was also used to examine the 40-year-old climate forecast by Exxon and how the concept of the individual carbon footprint originated from the PR of British Petroleum. Within the lecture, students were asked to consider if and how they have internalized a logic that foregrounds individual responsibility. In this iteration of the course, the students were also introduced to 'greenwashing'. Greenwashing is defined as '*poor environmental performance and positive communication about environmental performance*' or '*the act of misleading consumers regarding the environmental practices of a company or the environmental benefits of a product or service*' (Delmas and Burbano 2011: 67). These definitions capture how greenwashing is used to hide negative environmental impacts. After being introduced to the concept of greenwashing, students were asked if they had ever been manipulated by it. The purpose of this activity was to make it easier for students to identify discursive manipulations in order to avoid them in the future.

Following the interactive greenwashing activity, the course linked accessible and general arguments presented above to more rigorous research. It discussed carbon offsets and the limitations (or purposeful aim) of this approach (Acosta 2011; Böhm et al. 2012; Sherman 2013) and took a closer look at the limited impact of COVID lockdowns on global carbon emissions (Le Quéré et al. 2020; Tollefson 2021). The section ended with a discussion on how to productively tackle climate change, considering ongoing efforts to stall progress or even useful discursive framing of the issue. The discussion ended by posing the question 'Do we need to learn more to address global issues? Or do we have solutions but simply are not implementing them?', which set up the next two course parts where students met expert researchers with cutting edge knowledge about issues related to sustainability.

Part Two: Sustainability Challenges in Fisheries

The second part of the course changed gear and focused on complex sustainability challenges in natural resource extraction, focusing on fisheries and the variety of different factors affecting and complicating their management. The starting point was a global overview of fishery sustainability issues from industrial fisheries to small-scale fisheries and continued with an in-depth analysis using small-scale fisheries in East Africa as an example. The lecture was based on empirical data and personal experiences from relevant fieldwork and research. It described the challenges local fishing families and communities face, such as the high dependency on coastal resources, increasing fisher numbers, high fishing pressures, food-web changes, dwindling resources, the use of destructive fishing gears, and climate change impacts. Here the lecture takes up 'wicked problems' that small-scale fisheries and their management have to deal with in a context of extensive poverty while lacking livelihood alternatives. Further, global environmental change and unsustainable practices which may exacerbate inequalities were discussed and contextualized more broadly by identifying factors affecting these fisheries on higher levels and larger scales, such as lacking political will, corruption, lack of resources for monitoring, global market dynamics, foreign industrial fishing fleets, tourism, and the economic exploitation of developing areas. The lecture offered several starting points to think through possible means to increase social-ecological sustainability and encouraged students to develop, analyze, and rate different approaches.

Practical examples in this lecture stemmed from empirical data collection and fieldwork on small-scale fisheries in East Africa and fishers' perspectives on pathways to sustainability (for examples, see Wallner Hahn et al. 2018). This part starts off with pictures of fishers and fishing villages in East Africa, introducing students to their culture, practices, socio-economic context and struggles (Fig. 2.1).

One example focused on the use of destructive fishing gears, such as beach seines, which can harm fish producing habitats, leading to fewer fish and ultimately can result in poverty traps. It was then discussed how top-down regulative approaches were not adequate to solve this problem as they might have devastating socio-economic consequences for the fishers and their families, and how the focus rather should lie on cultural, normative and economic aspects, which could make it possible for fishers to change to more sustainable gears and potentially improve their livelihoods (Wallner Hahn et al. 2016). Other examples analyzed several well-intended internationally funded development cooperation projects in the area, which were unsuccessful or even increased negative socio-economic conditions and injustice because they were carried out inadequately or affected by corruption issues.

Part Three: Aquaculture, Sustainable Food Production, and the Intersection of Economics and Politics

The third course section mirrored the second but investigated aquaculture and food production as a case study. The logic of developing system-critical approaches from the bottom-up based on the rigorous study of real-life cases was the same. However, in contrast to the lecture in part two of the course, this lecture focused on food production in the developed world. The lecture began by introducing students to the various types of aquaculture, including their benefits, challenges, and trade-offs, with a focus on sustainability. The lecture highlighted how, although aquaculture is heralded as a potential solution for sustainable food production, its level of development in many countries is insufficient to realize said potential. The lecture then gave an overview of how, although aquaculture could be an important component of achieving sustainable food production, due to negative social perceptions, lack of education, and regulatory hurdles, aquaculture development in many developed countries is lacking. Leading to a situation where aquaculture development has effectively been outsourced to Asia where social acceptance is higher (Fig. 2.2), resulting in intensive production systems, which put pressure on their ecosystems, undermining their sustainability. As before, this course section aimed to provide students with concrete examples of possible means to tackle sustainability-related issues. Students were encouraged to think critically about the topic and questioned whether it is a lack of means or a lack of will that prevents developments to improve sustainability.

Fig. 2.1 (clockwise from top-left): Sieglind Wallner Hahn, *Scenes from fishing villages in Madagascar and Zanzibar, Tanzania*, 2015, Photographs. Copyright © Sieglind Wallner Hahn

Part Four: Students Rejecting Distractions and Looking for Real Solutions

The fourth and final course section aimed to bring the previous perspectives together. In a seminar called 'Ways forward', students presented what they considered a discursive distraction before suggesting a more meaningful alternative measure. Based on their own readings and previous lectures and discussions, they compiled a list of requirements that they believed any initiative towards sustainability should meet to avoid being

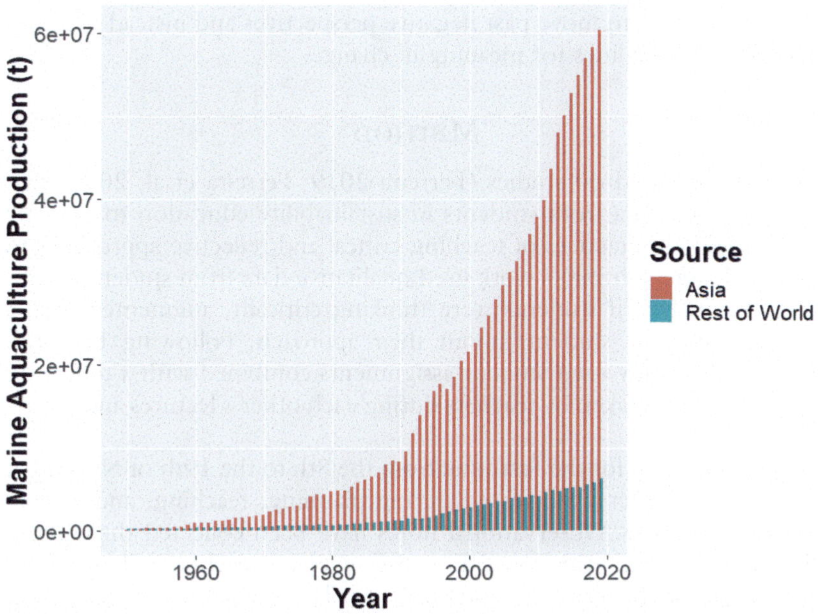

Fig. 2.2 *Marine aquaculture production (tons) in Asia compared to the rest of the world. Production values are based on the Food and Agriculture Organization of the United Nations global aquaculture statistics* (FAO 2022; https://www.fao.org/fishery/statistics-query/en/aquaculture)

considered deficient or inadequate. This task explicitly linked the critique of the dominant discursive frame of individualization to looking for actionable alternatives. While students were not expected to find actual solutions to global problems, linking alternative approaches to practical changes was meant to inspire critical thinking and problem solving, and ultimately encourage students to be more active and critical in discussions of global issues and avoid falling prey to discursive distractions. Students presented their criteria, explained their reasoning, and gave an example for an initiative that would meet them. The advantage of this approach was that it not only encouraged students to think critically about discussions around sustainability but also challenged them to consider which actions mattered or made an impact, a skill they could carry forward and apply in future dealings with sustainability issues. During discussions, students were encouraged to move past negative perspectives and instead focus on identifying possibilities for meaningful change.

Method

This work is based on studies (Ferreira 2019; Ferreira et al. 2013) that have collected data from students in sustainability education to evaluate interdisciplinary practices of teaching critical and reflective approaches to climate change. Ferreira´s work used qualitative data from student assignments to analyze if students were thinking critically, augmented with a survey asking the students about their approach. Following Ferreira's method, this study used student assignments combined with a follow up survey and notes taken by teachers during each other´s lectures and course planning.

The course ran for the first time from the 8th to the 19th of November 2021. This chapter was created during planning, teaching, and evaluation of the course. Observational notes have been collected throughout all phases and with the consent of the involved teachers (all of whom are authors of this chapter) and students. The evaluation of the course module is also based on student contributions and was shown to the students as a second way of ensuring that the analysis and conclusions are valid (Brown and Heller 1981).

Evaluation of the Course

Student Presentations

This section analyzes the presentations of students to evaluate if the course enabled them to investigate sustainability issues and think critically. During student presentations, unintended topics and themes that emerged were also discussed. The most remarkable element was the personal nature of the case studies chosen and the discussion of personal stories, family relationships, and their feelings in relation to sustainability.

The first topic in student presentations was critical thinking and scrutinizing to identify greenwashing. Students discussed the use of green images on products as well as labels like '*all-natural*' and '*eco-friendly*' that suggest they are good for the environment without really explaining why and how. They connected the slogan '*30% less climate impact*' to a marketing philosophy described as '*Buy green, buy more!*' and their conclusion here was that, as a consumer, it was difficult to see if any initiative was actually helpful or just marketing.

The second presentation was about how a coffee house chain deceived consumers to believe they dealt with waste disposal in a sustainable way. While the coffee house pretended to sort and recycle their waste, providing separate bins for different kinds of waste, in reality, everything went into the same dumpster behind the building. One of the students contacted a friend working for this coffee house, who confirmed the story:

> Both bags go straight to the trash at the end of the night! I asked about this as well when I first started working here. [...] Don't tell my name. I am not supposed to be telling you this. [...] Telling this is the crime. (student)

Students illustrated the limitations of 'right' consumer choices towards more sustainable products in terms of limited sustainable options (e.g., there is no relevant recycling for plastic), different economic preconditions and how consumers can be misled by producers or companies striving for profit rather than sustainability.

> I heard a quote one time that was: You vote with your dollar. But now I have heard so much that I don't think that people can know. I mean, what should I buy? (student)

Another student case study presented the case of a company their family members worked at delivering flower bouquets to supermarkets. In order to be able to use eco-friendly labels and consequently increase sales, the company switched to biodiesel for transporting flowers. The company then claimed that their products were eco-friendly, misleading consumers to believe that even the environmental impact of their flower production had been improved. This showcased for students why the power of consumers to achieve reduced environmental impacts of products is limited and that stronger demands on the responsibilities of producers could be more impactful.

Student Reflections

From in-class observations, students seemed interested and engaged in the course content. They actively participated in discussions, contributed ideas, and asked questions during lectures. Some of the students expressed after the course that they thought it was very interesting and that they really enjoyed it, as well as that it helped them to 'think outside the box'.

Students filling in the online survey had not taken other academic courses within the field of sustainability before. They reported that this course module contributed to new knowledge, widened their horizon on sustainability issues, and positively affected their critical thinking ability. Further, they felt this course module would be useful for their future career pathways.

Discussion

This course aimed to teach students to recognize when sustainability issues are being framed in a way that predominantly individualized responsibility, to be able to reject this frame and find systemic approaches to these questions instead. Based on observations during the course, analysis of student presentations, and the feedback from students in a post-course survey, it appears the course aim was achieved in some capacity, but also that there is room for improvement and that there are elements to critical sustainability education that were not sufficiently accounted for. Both observations during the course and the student presentations showed that students moved towards systemic perspectives, but quickly defaulted back to their known and familiar frames of individualized responsibility. In discussions during lectures, students frequently returned to consider individual diets, clothing choices, or travel habits. Necessitating prompts by the teacher to reconsider a different, systemic perspective again. The student presentations started with perspectives on greenwashing that, while centered around seeing through manipulation, are still done from the perspective of an, albeit informed and critical, customer. That said, the examples provided by students also considered systemic issues and showed an appreciation of sustainability issues as a symptom of systemic dysfunction. When asked in the course, students also mentioned that they had started to change their minds and had moved away from seeing the questions around sustainability as issues of personal virtue and responsibility. However, in the survey, they expressed that the different frames for sustainability questions and their role were not completely clear to them after the course. This shows that the course did not fully change their minds or get them to see the world through a critical and systemic lens. Which may have been too much to expect from a two-week course module in the face of life-long cultural hegemony. However, the critical examination of the work of their own families and the investigations of corporate practices with the help of friends indicates that the course at least gave the students a starting point for critical thinking and for seeing issues around sustainability from a systemic perspective.

The first iteration of this course was a test run with only five students and there may be limitations with larger class sizes. For example, having space for personal meetings with students and dealing with emotions. However, the small class size made it possible to have deeper conversations with each student and receive feedback.

To strengthen future iterations of this course module, we recommend a more stepwise process to engage students in critical thinking about sustainability issues. In the current iteration, we introduced students to several well-studied sustainability narratives, providing both the narrative and an explanation of how the narrative had been shaped and manipulated. Afterward, we challenged the students to select their own sustainability narrative and assess it critically. In future iterations, as an added activity, students should be given a set of well-studied sustainability narratives, but without information on how the narrative had been manipulated. Students would then be asked to assess whether they believe the narrative was manipulative in any way, and if so how, and asked to highlight what they believe to be strengths and weaknesses of the argument. After presenting their assessments of the narratives, instructors will reveal whether and how each narrative had or had not been manipulated. Through this activity, students will be given a chance to apply the skills and logic provided throughout the course module. At the same time, this activity will assess whether the course module had adequately prepared students to use critical thinking to navigate through complicated sustainability narratives.

During the lectures of this course module, ethical principles were touched upon, such as social justice concepts of equitable distribution, the precautionary principle, equity and justice or scientific knowledge, and integrity in decision making (UNESCO 2017). Values and social norms have been shown to influence behavioral intentions, and an ethical foundation should be a critical component of sustainability education (see also Biedenweg et al. 2010). Therefore, it would be vital for future courses to dedicate more lecture time to ethics in sustainability, both in theory and with practical applications such as student exercises.

Interdisciplinarity and Science Communication

There were values and limitations associated with the course's interdisciplinary nature. This course module was co-developed and taught by three teachers, who were present during most course modules together, and actively contributed to each other´s lectures, highlighting connections between topics and emphasizing the overall purpose of the course. It also required teachers to understand each other and their respective material, which was necessary to create lectures that built off one another, maintained a consistent message for students, and stimulated critical thinking. This level of teacher engagement necessitated a large amount of teaching hours which can limit implementation in some settings. However, creating space for teachers to interact across disciplines, learn from each other, and build bridges was useful, rewarding, and fun. The course also provided an opportunity for teachers to develop and expand their perspectives and practice communicating across disciplines.

Conflicting views about science communication also emerged during the course. Although there are calls for more passionate and provocative science communication to better engage society, many scientists find this sort of communication difficult and unsettling, preferring more carefully phrased and worded explanations. While careful, precise wording to avoid overstating findings makes sense in an academic context, in broader societal conversations such careful and deliberate phrasing can at times undermine confidence and risk not getting the message across (Brysse et al. 2013). While this is a known issue of science communication (Bickford et al. 2012), in interdisciplinary university education, this meant that some experts needed encouragement to speak more generally and use their expertise to support normative statements. Learning about this issue was also something new and surprising for the students:

> I always thought that they (the scientists) are so vague. Like: This could be connected... I never thought about that they are trained to do that. (student)

Another example concerning science communication that emerged during the course concerned the discussion of a particularly destructive fishing technique. Bottom trawling is a highly controversial fishing method, which has many negative environmental impacts (Hiddink et al. 2017), and there have been calls for more sustainable fishing methods

(Pusceddu et al. 2014; Shester and Ayers 2005). However, bottom trawling remains widely used and continues to negatively affect seabeds (Hiddink et al. 2017). During the course, students were surprised by this and questioned why a known destructive fishing technique continued to be used. This provided a useful example for teachers to highlight an area where political action has failed to follow known scientific advice. The point of the lesson was to exemplify that decision-makers did not always follow up on scientific advice. For this chapter, this situation showcases the advantage of interdisciplinary teaching: The expertise of scientists gets to inform a critical perspective on broader societal issues, the societal critique gains credibility and a connection to material reality. Here the two parts came together to form more than the sum of their pieces.

In this course, the setup of one perspective being more based on ethnographic work and grounded in practical examples along with a more systemic approach was especially useful. It made it possible to follow the argument for the need of systemic change from the discursive to the material and from the general and global to the specific and local. Ethnographic work, especially when done from a nature-scientific perspective, is a useful bridge between social sciences and nature science. It shows the human consequences of systemic changes and stresses how interconnected humans and nature are (even in an anthropocentric world view). The students could see, directly next to each other, that change in the way we discuss sustainability is needed. They could arrive at that conclusion based on analyzing the discursive framing of sustainability. They could come there from the starting point of the problems with small-scale fishing. They could see it when learning about the barriers for implementing working and sustainable methods for aquaculture. The combination of social and nature science, the global and the local, into the argument for a shared systemic approach discussing culture, but grounded in the material world, made the course moment work.

These different angles could be useful to keep for future course iterations and should perhaps be purposely planned for. Introducing additional perspectives could also be valuable and should be explored in future course iterations.

Questions of Shame and Guilt and the Need for Emotional Support

An unexpected theme that emerged during the course was the experience of shame and guilt. During breaks between lectures and in discussions of student presentations, students shared their experiences of shame and guilt. One student mentioned being criticized by their father for caring about sustainability. Another student explained that his generation had been blamed for climate change by their teacher in school as a response to their participation in climate change protests.

> I went to 'Fridays for the future' and the teacher let us go but she also then said that it is the fault of the younger people. She said that we are charging the phone many times a day and they are not doing that so it is the young people's fault. (student)
> I feel ashamed that I am taking this course because it is not something that is good or that it is going to affect… So when I am talking to people I say that I am taking this course and I say that I want to have it in fashion for later. But it is so connected to earlier, to the political movements in the 80s and… I feel that people think that it is connected to communism. (student)

The emergent topic of personal guilt and shame in relation to sustainability indicates that there is a need to address this in education. While this course started with a discussion of personal responsibility and virtue as the frames through which sustainability issues have been presented in mainstream culture, our intent was to make students aware of this framing and stimulate them to assess and question these frames critically. Framing climate change as an issue of personal character and ethical virtue instead of real and tangible changes in our economic- and production systems seems to have the consequence that even in an educational moment, feelings of shame push themselves into the conversation and need to be addressed. During the final seminar of the course, guilt and shame came up and took over part of the discussion. Guilt and shame were not topics the course was prepared for, nor was it prepared to handle their emotional impact on students. Such emotional responses have been recognized to be a relevant aspect of climate change mobilization (Kleres and Wettergren 2017).

The next version of this course module should consciously and explicitly address how the discursive frame also operates on shame. It could be possible to give the students the task to reflect on how their own thinking has been influenced and to help them see how they have been shamed so that they can grow beyond this approach. That said, it might be naïve to expect students to think themselves out of an emotional and personal issue like this and it could be worth considering a different approach to specifically address these emotional issues. There is an element of vulnerability here, as the sources of shaming reported in this course have been parents and teachers who should have been offering a safe space and support. As it stands, the course has not been specifically prepared for addressing this and will include this area in next year's planning.

Conclusion

The course module described in this chapter aimed to explore how interdisciplinary education at a university level can teach students to think critically about sustainability issues and their discursive framing. Specifically, the aim was to help students step beyond individualizing approaches to instead consider systemic perspectives on sustainability. Although this was only a small pilot course with five students, based on the analysis of the student assignments, notes taken during lectures, as well as a brief student survey, the course was successful in stimulating critical thinking to an extent. During their presentations, students questioned the effectiveness of sustainability actions, analyzed what impacts actions had, compared how impacts measured up to the goals they were meant to accomplish, and considered how sustainability actions were presented to consumers. Students reported in survey results that the course broadened their understanding of sustainability issues and had a positive effect on their ability to think critically.

Although there were challenges associated with the interdisciplinary nature of the pilot course, the benefits outweighed the cost. Connecting a critical frame perspective with expert analysis and case studies gave students a broad and holistic view of sustainability. In particular, combining an ethnographic case study with an analysis of more extensive systems is recommended. While these approaches required significant time investment on the part of the teachers, they were also rewarding for both students and teachers. Although such an intensive and time-consuming approach may limit class size, we recommend its use.

A final unexpected finding from this pilot course was the centrality of guilt and shame for students studying sustainability. Dealing with the emotional response of students to feelings of personalized responsibility towards sustainability was not something the course was designed for. As a result, future iterations of the course should be prepared to address these issues and perhaps consider broadening the interdisciplinary approach to incorporate experts in psychology and human behavior.

References

Acosta, Ines (2011), '"Green desert" monoculture forests spreading in Africa and South America', *The Guardian*, 16 September 2011, https://www.theguardian.com/environment/2011/sep/26/monoculture-forests-africa-south-america. Accessed 8 April 2022.

Andersen, Robin (2013), The 'Crying Indian,' Corporations, and Environmentalism A Half Century of Struggle over Environmental Messaging. Routledge Companion to Advertising and Promotional Culture.

Bickford, David, Posa, Mary Rose C., Qie, Lan, Campos-Arceiz, Ahimsa, and Kudavidanage, Enoka P. (2012), 'Science communication for biodiversity conservation'. *Biological Conservation*, 151(1), 74–76.

Biedenweg, Kelly, Monroe, Martha C. and Oxarart, Annie (2010), 'The importance of teaching ethics of sustainability', *International Journal of Sustainability in Higher Education*, 14 (1), 6–14.

Böhm, Steffen, Misoczky, Maria Ceci, and Moog, Sandra (2012), 'Greening Capitalism? A Marxist Critique of Carbon Markets', *Organization Studies*, 33(11), 1617–1638.

Brown, Alan, and Heller, Frank. (1981), 'Usefulness of group feedback analysis as a research method: Its application to a questionnaire study', *Human Relations*, 34(2), 141–156.

Brysse, Keynyn, Oreskes, Naomi, O'Reilly, Jessica, and Oppenheimer, Michael (2013), 'Climate change prediction: Erring on the side of least drama?' *Global Environmental Change*, 23(1), 327–337.

Delmas, Magali A., and Burbano, Vanessa Cuerel (2011), 'The drivers of greenwashing', *California Management Review*, 54(1), 64–87.

Environment, U. N. (2021), Making Peace With Nature. UNEP - UN Environment Programme, http://www.unep.org/resources/making-peace-nature

Food and Agriculture Organization of the United Nations (2022), *Fisheries and aquaculture production*. https://www.fao.org/fishery/statistics-query/en/aquaculture

Ferreira, Jo-Anne (2019), 'The limits of environmental educators' fashioning of 'individualized' environmental citizens', *The Journal of Environmental Education*, 50(4–6), 321–331.

Ferreira, Jo-Anne, Keliher, Vicky, and Blomfield, Jessica (2013), 'Becoming a reflective environmental educator: Students' insights on the benefits of reflective practice', *Reflective Practice*, 14(3), 368–380.

Folke, Carl, Polasky, Stephen, Rockström, Johan, Galaz, Victor, Westley, Frances, Lamont, Michele, Scheffer, Marten, Österblom, Henrik, Carpenter, Stephen

R., Chapin III, F. Stewart, Seto, Karen C., Weber, Elke U., Crona, Beatrice I., Daily, Gretchen C., Dasgupta, P.artha, Gaffney, Owen, Gordon, Line J., Hoff, Holger, Levin, Simon A., Lubchenco, Jane, Steffen, Will and Walker, Brian H. (2021), 'Our future in the Anthropocene biosphere', *Ambio*, 50, 834–869.

Hiddink, Jan Geert, Jennings, Simon, Sciberras, Marija, Szostek Claire L., Hughes, Kathryn M., Ellis, Nick, Rijnsdorp, Adriaan D., McConnaughey, Robert A., Mazor, Tessa, Hilborn, Ray, Collie, Jeremy S., Pitcher, C. Roland, Amoroso, Ricardo O., Parma, Ana M., Suuronen, Petri and Kaiser, Michel J. (2017), 'Global analysis of depletion and recovery of seabed biota after bottom trawling disturbance', *Proceedings of the National Academy of Sciences*, 114(31), 8301–8306.

Kent, Jennifer (2009), 'Individualized responsibility and climate change: "If climate protection becomes everyone's responsibility, does it end up being no-one's?"', *Cosmopolitan Civil Societies: An Interdisciplinary Journal*, 1(3), 132–149.

Kleres, Jochen, and Wettergren, Åsa (2017), 'Fear, hope, anger, and guilt in climate activism', *Social Movement Studies*, 16(5), 507–519.

Laufer, William S. (2003), 'Social accountability and corporate greenwashing', *Journal of Business Ethics*, 43(3), 253–261.

Le Quéré, Corinne, Jackson, Robert B., Jones, Matthem W., Smith, Adam J. P., Abernethy, Sam, Andrew, Robbie M., De-Gol, A.nthony J., Willis, David R., Shan, Yuli, Canadell, Josep G., Friedlingstein, Pierre, Creutzig, Felix, and Peters, Glen P. (2020), 'Temporary reduction in daily global CO2 emissions during the COVID-19 forced confinement', *Nature Climate Change*, 10(7), 647–653.

Lerner, Sharon (2019), 'How the Plastics Industry Is Fighting to Keep Polluting the World', *The Intercept*. 20 July 2019, https://theintercept.com/2019/07/20/plastics-industry-plastic-recycling/. Retrieved April 8, 2022

Maniates, Michael F. (2001), 'Individualization: Plant a tree, buy a bike, save the world?' *Global Environmental Politics*, 1(3), 31–52.

Pusceddu, Antonio, Bianchelli, Silvia, Martín, Jacobo, Puig, Pere, (2014), 'Chronic and intensive bottom trawling impairs deep-sea biodiversity and ecosystem functioning', *Proc Natl Acad Sci*. 111(24):8861–8866.

Sherman, Daniel J. (2013), 'Upsetting the Offset: The Political Economy of Carbon Markets', *Social Movement Studies*, 12(3), 354–355.

Shester, Geoff, Ayers, Jim (2005), 'A cost effective approach to protecting deep-sea coral and sponge ecosystems with an application to Alaska's Aleutian Islands region. In: Freiwald, A., Roberts, J.M. (eds) *Cold-Water Corals and Ecosystems*. Erlangen Earth Conference Series. Springer, Berlin, Heidelberg.

Supran, Geoffrey, and Oreskes, Naomi (2021), 'Rhetoric and frame analysis of ExxonMobil's climate change communications', *One Earth*, 4(5), 696–719.

Tollefson, Jeff (2021), 'COVID curbed carbon emissions in 2020—But not by much', *Nature*, 589(7842), 343–343.

UNESCO (2017), 'Declaration of Ethical Principles'. https://en.unesco.org/themes/ethics-science-and-technology/ethical-principles, Accessed 08 April 2022.

UNESCO (2019), 'Framework for the implementation of education for sustainable development (ESD) beyond 2019', *UNESCO General Conference*, 40th, Paris, 2019, 40 C/23.

Wallner Hahn, Sieglind, Molander, Fia, Gallardo, Gloria, Villasante, Sebastian, Eköf, J. Ohan S., Jiddawi, Narriman S. andde la Torre-Castro, Maricela (2016), 'Destructive gear use in a tropical fishery: Institutional factors influencing the willingness- and capacity to change', *Marine Policy* 72, 199–210.

Wallner Hahn, Sieglind, de la Torre-Castro, Maricela (2018), 'Early steps for successful management in small-scale fisheries: An analysis of fishers', managers' and scientists' opinions preceding implementation', *Marine Pollution Bulletin*, 134, 186–196.

Wals, A. E. (2019), 'Sustainability-oriented ecologies of learning: A response to systemic global dysfunction', in *Ecologies for Learning and Practice*, Routledge, pp. 61–78, https://www.taylorfrancis.com/chapters/edit/10.4324/9781351020268-5/sustainability-oriented-ecologies-learning-arjen-wals.

CHAPTER 3

Greta Thunberg and the Body of Politics: How Young Climate Change Protesters are Reminding Us of the Gift of Materiality

Francesco Buscemi

Introduction

This chapter semiotically analyses the use of the body and the resort to the materiality of the young protesters against climate change, specifically Greta Thunberg and the Fridays for Future demonstrators. The study also confronts the semiotics of the body of these protesters to the digital immateriality put forward by virtual reality and the supposed metaverse theorized by Facebook's co-founder and chairman Mark Zuckerberg, which in the last few years has frequently been pointed out as key to the future of politics and of the human relationships in general.

F. Buscemi (✉)
Department of Communication and Performing Arts, Catholic University of Milan, Milan, Italy
e-mail: francesco.buscemi@unicatt.it

D. Binns and R. Najdowski (eds.), *Confronting the Climate Crisis*, Palgrave Studies in Media and Environmental Communication,
https://doi.org/10.1007/978-3-031-89606-4_3

Greta Thunberg and the protesters belonging to the Fridays for Future movement are significant representatives of the new environmental advocates speaking out all over the world against climate change and the absence of effective environmental policies in the industrialized nations. Thunberg became globally popular in the summer 2018 when, at 15, she started protesting in front of the Swedish Parliament to persuade Swedish politicians to become aware of global warming and change their environmental policies (Kraemer 2021).

Thunberg would come back to that place almost each Friday for more than one year, in sun, rain, snow and ice. The protest spread over the world, involving other people, mostly youngsters, and on 15 March 2019, it became global with 71 counties and more than 700 places asking all the politicians on the planet to intervene to avoid the imminent disaster caused by the rise of temperature (Watts 2019). On 23 September 2019, Thunberg *addressed the UN's Climate Action Summit. 'You have stolen my dreams and education (…) and all you can talk about is money and fairy tales of eternal economic growth' (*Cooper 2019*), she said to a very powerful audience, even though Scott Morrison, Prime Minister of Australia and the Brazilian leader Jair Bolsonaro resulted missing (*Cooper 2019*). From that moment onwards, Thunberg met some of the most important political leaders internationally and her words have never changed so far.*

Fridays for Future is the global movement generated by Thunberg's protest. As the Swedish campaigner used to stay in front of the Swedish Parliament every Friday, they decided to gather on the same day in many cities to fight the passivity of the politicians in front of the urgency of climate change. As reported by their website, the movement is today present in 7500 cities on all continents and involves more than 14 million people (Fridays for Future 2022). The pandemic seemed to mean the end of the movement, as gathering in public places was forbidden in the majority of the countries where the youngsters used to protest. However, during the various lockdown restrictions, the movement survived online and as the measures became more permissive, the Fridays of the major cities were filled with young students and some adults asking for more attention to the global warming problems again (Mlaba 2021).

Existing Literature

This section reviews the literature on the young protesters against climate change and on virtual politics, as these are the two fields that I combine in this study. Greta Thunberg and the Fridays for Future movement have often been analysed as two examples challenging the supposed disinterest of the youngsters in politics, as has been advanced by many scholars. Bowman (2019) points out that the issue of young protesters against climate change has been investigated from a strict binary point of view, separating public engagement and private emotions. Conversely, these protests tend to merge the two levels and offer a new way of protesting. Majeed and Lee (2017) advance that the youngsters show much interest in global warming because of the awareness that this problem affects their lives strongly, even increasing uncertainty and causing depression and mental health issues in their minds.

Bowman et al. (2021) argue that research on young protesters against global warming is written by adults and is adult-centred. This means that in many studies the young protesters are pigeonholed into categories created by adults that the new generations cannot comprehend. Moreover, adults tend to classify them as politically passive just because they express themselves through behaviours that the adults do not consider political forms of protest. Thomas et al. (2019) find that these forms of protest have taken various modalities according to the place where they have occurred.

Specifically on Thunberg, and importantly for this study, Kraftl (2020) notes that the Swedish campaigner's discourse is based on action rather than on hope. This shifts the balance of her protest towards a concrete answer, confining more abstract dreams of a better world to the background. Peters (2020) argues that Thunberg and the young protesters of the Fridays for Future movement are a novelty for western culture, as in this context young people have almost never been represented as bearers of hope and future. Finally, Jackson (2021) links Thunberg's protest to the materiality of the world. He in fact argues that Thunberg reminds us that the world is made of materiality and that we are destroying it.

Certainly, both Thunberg and the other young protesters have also adopted virtual, immaterial media to promote their ideas. On virtual media and politics, some theorists and practitioners have seemed enthusiastic about the advantages that virtual technologies may bring to politics, while others show doubts especially because of the scarce virtual media literacy around the world. Willson (1997) underlines that one of the most revolutionary innovations of virtual politics will be the virtual community, which for some will give participants a sense of liberation from the old form of physical participation; however, he continues, other scholars underrate virtual politics, foreseeing that the physical connection will not disappear, because of the limits of the virtual relationships.

Digital immateriality in the social context has risen as a hot topic when Mark Zuckerberg, Facebook's co-founder and chairman, has announced the metaverse, a sort of future development of Facebook in particular and social media more broadly. It consists of a virtual space where any user may virtually move to other places in order to meet other people, either in a work context or in their private lives. In short, the metaverse is a VR environment where various humans interact with each other (Bolter et al. 2021) and could mean going along with a friend to virtually attend a football match on the other side of the world with many other people connected from other places; or virtually having a work meeting with colleagues at the office.

Baker (2022) shows his fear of the contribution of the metaverse to politics. Politicians have always created false scenarios where they are depicted as perfect and in the right position to save the planet. Actually, the new immersive media may facilitate the production of these fictitious worlds; social media are already an example of this, as they create imaginary worlds apparently real, and this scenario may worsen because of the even more powerful metaverse.

Critically, McCourt (2005) sustains that the digital and immaterial media cannot contain their own history and that only having a concrete and material surface guarantees awareness and depth. Similarly, Barassi (2015) underlines the necessity of materiality in activist media.

Theories on Materiality and Immateriality

Electronic media such as TV and the internet have gradually dematerialized almost every step of the process leading from production to reception. Immateriality seemed to be, therefore, the new nature of media messages.

Everything starts from Baudrillard (1994), who argues that the media have covered the materiality of the real world with a veil of immaterial representations which hides the truth and re-creates reality. The result is a new world that is certainly unreal but that is also the unique one that people may perceive, so for all of them it becomes hyperreal, a world more real than reality.

Such a trenchant view has been adapted to the gradually more invasive digital media, which today shape our world. Lyotard (1996) highlights that the process leading to the de-materialization of the message is not a technological issue. Rather, it is humans, with their knowledge, who tend to de-materialize what they know. Specifically, Lyotard (1996) advances that semiotics and its notion of sign tend to substitute reality, replacing the real world with an immaterial simulacrum to be understood. Genosko (2016) objects that signs are already in the world and semiotics only decodes them. Materiality is thus around us and Parikka (2007) points out that even the digital has something material, such as its concrete disease, for example viruses. He thus considers the digital as a material body, an example of the materiality of culture.

In this contended context, the metaverse has seemed to demonstrate the close-future domination of immateriality. Saker and Frith (2022) underline the huge power of this immaterial technology. This raises questions about the identity of the users. Are they the same people as when they are out of the metaverse? Saker and Frith (2022) also show fear as such a powerful technology is in the hands of a company that has proven uneven ethical principles.

Coole and Frost (2010) find that, from the 1980s, the dominant discourse has focused on immaterial concepts such as language, culture or values and that it is time to go back to materiality for three reasons: firstly, the new developments in natural sciences which shift from human centrism to a more balanced posthumanism; secondly, biopolitics and bioethics, considering the status of the planet and its inhabitants; and finally, political economy, as geopolitical and socioeconomic structures occur through material processes.

In line with this idea of returning to materiality, Buscemi (2022) analyses the relationships between material and immaterial media within the same context and finds that immaterial media such as virtual reality does not give the audience pieces of information or knowledge. It only stupefies and offers the public emotions. On the contrary, material forms of expression considered old and surpassed such as theatre and television inform and increase people's knowledge. This means that immaterial digital media need the support of materiality to communicate complex content.

The inseparable co-existence of materiality and immateriality in the various forms of communication, from the political to the media, is explained by Biesecker and Lucaites (2009), when they say that in the end rhetoric is material. As rhetoric is the basis of any public discourse, political communication and media texts, this assumption has great importance for this study. Biesecker and Lucaites (2009) point out that rhetoric has meant different things and has had different aims over the years, from ancient Greece to Medieval Paris to Renaissance Florence, and that it has continued to change until the present. It has been immaterial for many of these ages, in the form of rules of composition or abstract norms about how to say something.

However, in the last few years technological and social developments have led rhetoric to a new change. It has become a form of experience created by the intersection of four concrete elements: speaker, speech, audience and event. Differently from the past, there is no one of them dominating the others generally, but each of them may be, at a time, in charge, and the final link of this chain is always change. In being forged by these forms of materiality, rhetoric has become material, not in the sense of something I can touch, but in the sense that it is a social and political force that impacts on the world (Biesecker and Lucaites 2009).

Starting from this wider idea of materiality, this study wants to understand how and why Greta Thunberg and the other young protesters against global warming take position between the opposite poles of materiality and immateriality, and what this means in comparison to the immateriality of the new media and the metaverse.

The Semiotics of the Body, a Method of Inquiry

In order to analyse Greta Thunberg and the Fridays for Future manifestations, this study starts from McGarry et al. (2020)idea that a protest is a form of aesthetics. Certainly, not a kind of aesthetics in Kantian terms, that is, an expression of pure beauty or perfection. Rather, a protest has an aesthetic value as it is first of all performance and confirms the theory that today aesthetics may push social change (Olcese and Savage 2015). As a form of aesthetic expression, thus, a protest may be analysed through semiotics. McGarry et al. (2020) maintain that protestors elaborate and communicate ideas through aesthetic choices and that they also do so through their bodies, along with writings, clothes, objects and other material things.

The body is thus one of the material instruments that protesters adopt to act, create aesthetic performances give birth to messages and communicate ideas. Their bodies, thus, produce meaning, are material elements that may be read as semiotic signs that signify something. Not by chance, semiotics considers the body as a sign that can be decoded like any other element of the semiosphere. Interestingly, the body certainly pertains to nature, but since Medieval times it has also become a cultural artefact, being represented as an icon, index or symbol and a part of the semiotic cultural system, or corposphere (Finol 2015). The corposphere is defined as "the set of languages that originate from, and are modernized and performed by the body, seen as a semiotic complex" (Finol 2015: 82). This body's semiotic essence becomes even clearer when the body is a part of rituals or social events. In these cases, the body becomes action, which becomes a symbolic system.

This is exactly how this study analyses the bodies of Greta Thunberg and the other young protesters, which are thus analysed as signs meaning something through iconicity, indexicality and symbolism. Peirce (1960) argues in fact that a sign may be an icon, an index or a symbol according to the relationships between the signifier and the signified. It is an icon when the signifier physically resembles what it represents, for example in the case of a portrait depicting a person. The sign is an index when the signifier is linked through causality or contiguity to something else; for instance, in the case of the smoke meaning that somewhere nearby there is a fire. Finally, a sign is a symbol when the signifier and signified are connected arbitrarily, in a relationship that is acknowledged within a community, as in the case of the red colour meaning passion, which is a recognised meaning in some groups but not in every part of the planet (Jenks 1998). Importantly, a sign has usually all these characteristics, but to different extents. Frequently, one of the three categories is dominant, while the other two are secondary, or one of them does not exist at all.

In order to analyse the bodies of the protesters more in depth, this study has sampled two specific manifestations by Greta Thunberg and another two by the Fridays for Future movement. They are the protest in front of the Swedish Parliament in 2018 and the speech at the UN Climate Action Summit on 23 September 2019 (including the trip to reach New York) for Thunberg; and the global protests on 15 March 2019 and of 24 September 2021 for the Fridays for Future movement.

They are the most important events that both Thunberg and the movement have organized, as underlined by various sources (de Moor et al. 2020; Kraemer 2021).

Greta Thunberg and the Others Between Materiality and Immateriality

Greta Thunberg

The first element emerging from Thunberg's protest in front of the Swedish Parliament is that it is a material answer to a material problem that however is perceived as immaterial. Climate change is, in fact, a material issue, its physical effects are in part already in front of us and in part will change our lives even in more material ways in the future, if we do not react. However, when scholars research why people tend to remain passive in front of this problem, their most frequent answer is that it is because the large majority of them do not perceive global warming as real, tangible or seeable (Tam 2018; Mortillaro 2019).

Thunberg's body in front of the Parliament was thus a version of that form of aesthetic performance (McGarry et al. 2020) that has been mentioned above and that may be analysed in its being a cultural item (Finol 2015) enacting an aesthetic text (McGarry et al. 2020) leading to change (Olcese and Savage 2015). The 15-year-old campaigner sitting on the side of the street, still, alone, only a small element close to the huge institutional building, was the perfect visualization of her smallness in front of the hugeness of the problem of global warming. Semiotically, we can consider it as a meaningful system of signs, agreeing with Genosko (2016) that signs are in our daily lives, and not in our theories. As it has been seen in Finol (2015), Thunberg's body has become action, even though still, as her staying in front of the Parliament has produced a global movement, growing consciousness and the rise of a new way of considering the planet and our role on it; continuing to agree with Finol (2015), the action has become a symbolic system, a semiotic whole of signs that we analyse in order to understand its relationships with materiality and immateriality.

The system of signs of Thunberg's body in front of the Parliament totally draws on materiality, or better the reaction of materiality in a world that risks becoming immaterial and that people perceive as immaterial (Baudrillard 1994), as the problem she is fighting. Semiotically,

Thunberg's sat on the pavement was clearly an icon easily recognizable; but the dominant nature of her protest was indexical. In fact, immediately after the first protests, the campaigner became an index, a system of signs immediately and logically linkable to the problem of climate change. Finally, she has never been a symbol, as the link between signifier and signified is not arbitrary here. Thunberg was fighting global warming and was there for that reason, so there was no arbitrary connection between her and the issue of global warming; rather, there was a concrete association between her and the meaning she was producing.

The same contrast between smallness and hugeness occurred in the case of the UN speech. The young campaigner was invited by the biggest political institution of the world, the United Nations, and in the most important political building on the planet, its headquarter in New York. Again, she brought her small body to warn the world, but this time she was not alone, but surrounded by the bodies of the most important political leaders internationally.

The difference from the other bodies, however, was in the way she reached New York. While the leaders did it by flight, increasing pollution, Thunberg travelled through the sea, sailing on the Malizia 2 from Plymouth in the UK to New York, and going back to Europe on the catamaran La Vagabonde from Hampton, Virginia, to Lisbon, in order to trip in a carbon-neutral way. While the most powerful people of the world fast appeared in and disappeared from New York, she showed the world the materiality (and the issues) of travelling, posting on her social media accounts images of her trip and the very material problems that travelling carbon neutral implies: Malizia 2, in fact, did not have toilets, cooking facilities, beds and showers as we know them. As previously, she stressed the materiality of these issues in her posts.

Again, it was smallness against bigness. A young campaigner on a small boat (along with other people, clearly) in comparison to the hugeness of the Atlantic Ocean. And as underlined by Finol (2015), her body became action, the solitary rebellion against the passive global leaders ignoring the close catastrophe; and the action became a semiotic system of signs: the fight between the smallness of the campaigner's body and the hugeness of the problem; the materiality of her actions and the immateriality of the perception of the problem; and even the materiality of the problems posted on social media and the immateriality of the leaders' disappearance from New York.

What is more, what has been said above in terms of the semiotic categories is confirmed here. Apart from the iconic nature of the sign (Thunberg onboard, for example), the indexical dominance is evident. Thunberg is immediately connectable to the fight against global warming, while any symbolical nuance of the sign results absent, as there are no arbitrary links but only concrete relationships keeping together the signifier and the signifying.

Finally, there is the issue of the immaterial media. Even Thunberg, in fact, abundantly uses social media. The trip on Malizia 2, for example, was told by the Swedish campaigner as a diary, often communicating the exact position of the ship in the Atlantic Ocean and keeping her followers updated on what happened aboard. Another Thunberg's use of social media is to share data or to reinforce concepts and theories to confirm what she has said during events or rallies. Her face and body are very often present in these messages.

What is clear, however, is that what Thunberg says on social media is in the background in comparison to what she really does. Posts, photos and diaries are always referred to real events and thus her immaterial life on social media seems to be a support of her material protest. Almost never, what she writes on Twitter ends on Twitter or other platforms. Being on social media, for her, is a preparation for something real that would happen soon or had happened before. Finally, as suggested by Bowman (2019) for the new protesters in general, Thunberg's use of social media mixes cognitive and emotional elements. The diary of the trip to New York, for example, is full of personal notes on her moods, impressions, fear, etc. This clearly gives these posts a more concrete perspective.

Fridays for Future

In the demonstrations of the Fridays for Future movement on 15 March 2019 and 24 September 2021, the chain body/action/semiotic system behaves differently. First of all, the body is actually *the bodies*, as the movement is formed by a huge quantity of demonstrators and not by a unique person. These many bodies produce an action that is made up of a movement flooding the squares and the streets all over the world. Thunberg's solitary stillness is not here, even though it produced the spark originating the movement. The resulting semiotic system is different from the previous, as this time, the hugeness belongs to the protesters, who are a crowd filling squares and streets that appear to be small in comparison to the big wave of the protest.

The aesthetics of this protest, in the terms advanced by Olcese and Savage (2015) and McGarry et al. (2020) and developed in the theoretical framework of this study, is a collective performance more joyful and optimistic than Thunberg's one. While the Swedish campaigner links more to anger than to hope (Kraftl 2020), and rarely smiles, the youngsters of the Fridays for Future movement smile, sing, shout and seem to never stop. Even though anger is present in their discourse, thus, hope seems to be the dominant sentiment, as pointed out by Peters (2020).

What is not lost in comparison to Thunberg is the resort to materiality. The people of the movement meet in person and put their bodies on the streets in the rain, snow or whatever may happen. The pandemic interrupted their manifestations only for a few months, and as soon as in one city it was possible to gather, they immediately organized a new rally. The urgency, thus, was to bring their bodies to say no to global warming.

The only semiotic similarity that remains with Thunberg's protest is their indexicality. The political discourse of the movement, in fact, refuses symbolism. They do not show symbols such as flowers, drawings, or other elements that may have arbitrary links to the content of their protests. Rather, their presence and slogans are directly associated to the fight against global warming. Even the logo they have chosen for their organization is purely indexical: it represents the world, with the recognizable continents, with the writing 'Fridays for Future'. No abstract images, no symbols but only the concrete and material link to the theme of their campaign. Even the green colour of the background is more indexical than symbolical. It is the colour of plants and other natural elements, which the movement aims to save.

As with Thunberg, the immaterial side of the Fridays for Future movement is on social media, and even more than in Thunberg's ones, it consists of a detailed account of the manifestations of the various national branches of the movement around the world. Again, social media are at reality's service. Immateriality, thus, is not seen as an independent dimension able to create its reality, but as a simple means through which the irreplaceable reality may be communicated and amplified.

Conclusion

We may say that thanks to Greta Thunberg and the other young protesters, the body and materiality are back. After Baudrillard's (1994) worrying theory that reality is disappearing and is going to be totally covered by immaterial media representations, the recent rallies against climate change are putting forward the opposite perspective. While Zuckerberg's metaverse seems to be the concrete development of Baudrillard's fears about immateriality and the absence of the body, Greta Thunberg and the young protesters stemming from her revolt are telling us that future politics needs to rely on body and materiality. Leaving to the old politics the ability to create fictitious worlds (Baker 2022) and to rapidly disappear from the UN building in New York, Thunberg and her disciples build on reality and materiality, even the materiality of a difficult trip across the Atlantic Ocean. This may usefully be linked to what is underlined by Jackson (2021) in the literature review, that this determined attachment to materiality has to do with the materiality of Earth, nature and the environment, all elements that these young campaigners want to save and that are central to their politics and, I would say, entire lives.

All of this shapes the contemporary form of rhetoric advanced by Biesecker and Lucaites (2009), which has become material thanks to the relationships among its main concrete elements: speaker, speech, audience and event. As prefigured by the two scholars, in this new material rhetoric, the main role among the four elements is not fixed, and each of the four elements may be in charge in turn. Not by chance, in Thunberg sitting in front of the parliament, the rhetoric is based on the speaker (even though she was mostly silent, but we are talking about her discourse in a wider sense); in the case of Thunberg addressing the UN assembly, the dominant trait of her rhetorical system was the speech; and finally, in the cases of the Fridays for Future manifestations, the principal element is certainly the event. Interestingly, it is the audience the missing element. This confirms that the messages produced in these manifestations still strive to find a deep reception from those who do not participate in the events and that there is still a gap to be filled. These are the ways that this contemporary, material rhetoric impacts the world, as in Biesecker and Lucaites (2009).

Moreover, if we can easily believe that the metaverse and every other distancing from the actual body have also been boosted by the fear caused by COVID-19, we can also say that this return to the body and materiality has occurred despite the pandemic, as is confirmed by the rallies organised by the Fridays for Future movement in various countries immediately after the restrictions were abolished. If we look at this from a generational perspective, we can see that, paradoxically, the youngsters, who for the adults superficially live in an immaterial world made up of social media, selfies, VR and similar artifices, are reminding us that materiality is a gift to be used.

Thunberg and the others' use of social media, in fact, suggests that the digital can only be a form of preparation for the real, and not the replacement of the actual world. The political virtual community prefigured by Willson (1997), thus, is only the preparation for the real community that is going to gather in a square, or the celebration of the actual one that manifested a few days earlier. This may be considered a new form of materiality of social media, adding to the one suggested by Parikka (2007). If social media representations refuse to be a world per se but support reality, their immateriality certainly persists, but also becomes more connected to the actual world. In another work (Buscemi 2022), I have demonstrated that the virtual needs the real when it comes to giving the audience non only stupor and marvel, but also information and knowledge. This research, thus, reinforces this theory.

This more realistic adoption of social media also secures a safer treatment of people's identity when they are online. Saker and Frith's (2022) fear of the creation of another identity when one is online in these cases tends to shrink. The way Thunberg and the other campaigners use these platforms, in fact, emphasises their real identity and does not ask for the creation of a new one. While this is easier in the case of a single protagonist, as in the case of Thunberg, this is more difficult when a collective of people, the Friday for Future protesters, are on the stage. However, even in this case, the continuous repeating of the images of their rallies around the world contributes to this attachment to reality.

Virtual politics has failed, these young demonstrators are telling us, and we need leaders putting their bodies in their concrete battles, rather than immaterial representations of people and ideas as suggested by the metaverse. This re-appropriation of the body may happen in two different ways: stressing the anger, as in the case of Thunberg and as suggested by Kraftl (2020); or highlighting a more joyful component of hope, like in the crowds of the Fridays for Future protesters. Both approaches, however, have managed to overcome the centrality of the immaterial, as suggested by the metaverse replacing reality.

To explain how the media have covered reality, Baudrillard (1994) starts from a Jorge Louis Borges's short story, where the Emperor asks a group of cartographers to realise the perfect map of the Empire. The various drafts are rejected by the Emperor, as they do not reproduce the actual territory. In the end, the cartographers succeed, realizing the perfect map, which reproduces each element of the real Empire entirely. However, this map is so big that it totally covers the territory. As a result, people look at the map believing it is the real land of the Empire. Baudrillard (1994) says that the media are doing the same, replacing reality and making people think that the media representations are reality.

Starting from this, it is as if Greta Thunberg and the other protesters had lifted Baudrillard's map and had let reality become visible again. They have shown us what is under the immaterial map of the metaverse, VR and the other technologies we are getting used to, to the point of considering as reality what they represent virtually.

Finally, following on from Peirce (1960), it is important to underline that these young protesters are mostly indices, sometimes icons (in the case of Thunberg), but never symbols, and this again reinforces their material status and their distance from immateriality. Icons directly resemble materiality; indices refer to something material that is absent but reminded by the indexical sign; while the symbol is totally immaterial, being the link between signifier and signified based on an arbitrary idea shared within a community. By personifying the protesters with the first two categories and rejecting the third, Thunberg and the others have further insisted on the side of materiality.

References

Baker, Gerard (2022), 'Politics are already leading us into the metaverse', *Wall Street Journal*, 24 January, https://www.wsj.com/articles/politics-already-leading-us-into-metaverse-schumer-biden-jan-6-insurrection-woke-crt-big-lie-11643040890. Accessed 29 March 2022.

Barassi, Veronica, (2015), *Activism on the Web: Everyday Struggles against Digital Capitalism,* New York: Routledge.

Baudrillard, Jean (1994), *Simulacra and Simulation,* Ann Arbor: The University of Michigan Press.

Biesecker, Barbara and Lucaites, John Louis (2009), 'Introduction', in Barbara Biesecker and John Louis Lacaites (eds), *Rhetoric, Materiality, & Politics*, New York: Peter Lang, pp. 1–16.

Bolter, Jay David, Engberg, Maria and Macintyre, Blair (2021), *Reality Media: Augmented and Virtual Reality*, Boston: The MIT Press.

Bowman, Benjamin (2019), 'Imagining future worlds alongside young climate activists: A new framework for research', *Fennia: International Journal of Geography*, 197:2, pp. 295–305.

Bowman, Benjamin, Bell, Karin and Alexis-Martin, Becky (2021), 'Youth, climate and environmentalism', in Karen Bell (ed.), *Diversity and Inclusion in Environmentalism*, London: Routledge, pp. 132–147.

Buscemi, Francesco (2022) 'The paradox of the virtual: Iñárritu's *Carne y Arena* between innovative spect-actor and traditional fruition', *New Techno-Humanities,* 2:1, pp. 23–27.

Coole, Diana and Frost, Samantha (2010), 'Introducing the new materialisms', in Diana Coole and Samantha Frost (eds), *New Materialisms: Ontology, Agency, and Politics*, Durham: Duke University Press, pp. 1–45.

Cooper, Rachel (2019), '"You have stolen my education": Greta Thunberg's powerful speech at the UN Climate Summit', *Climate Action*, 24 September, https://www.climateaction.org/news/you-have-stolen-my-education-greta-thunbergs-powerful-speech-at-the-un-clim. Accessed 23 March 2019.

de Moor, Jost, Uba, Katrin, Wahlström, Mattias, Wennerhag, Magnus and De Vydt, Michiel (2020), Protest for a future II: Composition, mobilization and motives of the participants in Fridays For Future climate protests on 20–27 September, 2019, in 19 cities around the world', report, https://researchportal.helsinki.fi/en/publications/protest-for-a-future-ii-composition-mobilization-and-motives-of-t. Accessed 29 March 2022.

Finol, José Enrique (2015), *On the Corposhere*, Berlin: Walter de Gruyter.

Fridays for Future (2022), 'What we do', *Fridays for Future*, 23 March, https://fridaysforfuture.org/. Accessed 23 March 2022.

Genosko, Gary (2016), *Critical Semiotics: Theory, from Information to Affect*, New York: Bloomsbury.

Jackson, Tim (2021) *Post Growth: Life after Capitalism.* Cambridge: Polity Press.

Jenks, Chris (1998), *Core Sociological Dichotomies,* London: Sage.

Kraemer, Daniel (2021), 'Greta Thunberg: Who is the climate campaigner and what are her aims', *BBC News*, 5 November, https://www.bbc.com/news/world-europe-49918719. Accessed 22 March 2022.

Kraftl, Peter, (2020), *After Childhood: Re-thinking Environment, Materiality and Media in Children's Lives*, Abingdon and New York: Routledge.

Lyotard, Jean François (1996), 'Les immateriaux', in Ressa Greenberg, Bruce W. Ferguson and Sandy Nairne (eds), *Thinking about Exhibition*, Abingdon: Routledge, pp. 159–175.

Majeed, Haris and Lee, Jonathan (2017), 'The impact of climate change on youth depression and mental health', *Lancet Planetary Health*, 1:3, pp. 94–95.

McCourt, Tom (2005), 'Collecting music in the digital realm', *Popular Music and Society*, 28:2, pp. 249–252.

McGarry, Aidan, Erhart, Itir, Eslen-Ziya, Hande, Jenzen, Olu and Korkut, Umut (2020), 'Introduction: The aesthetics of global protest: Visual culture and communication', in Aidan McGarry, Itir Erhart, Hande Eslen-Ziya, Olu Jenzen and Umut Korkut (eds), *The Aesthetics of Global Protest: Visual Culture and Communication*, Amsterdam: Amsterdam University Press, pp. 15–37.

Mlaba, Khanyi (2021), 'Fridays for Future: Young climate activists around the world take to the streets', *Global Citizen*, 24 September, https://www.globalcitizen.org/en/content/fridays-for-future-climate-protest-cop26/. Accessed 23 March 2022.

Mortillaro, Nicole (2019), 'The psychology of climate change: Why people deny the evidence', *CBC News*, 29 January, https://www.cbc.ca/news/science/climate-change-psychology-1.4920872. Accessed 27 March 2022.

Olcese, Cristiana and Savage, Mike (2015), 'Notes towards a "social aesthetics"', *British Journal of Sociology*, 66:4, pp. 720–737.

Parikka, Jussi (2007), *Digital Contagions: A Media Archaeology of Computer Viruses*, New York: Peter Lang.

Peirce, Charles S. (1960), *Collected Papers. Vol. II: Elements of Logic*, Cambridge: Harvard University Press.

Peters, Laura (2020), 'The limits of the human? Exhibiting colonial orphans in Victorian culture', in Diane Warren and Laura Peters (eds), *Rereading Orphanhood: Texts, Inheritance, Kin*, Edinburgh: Edinburgh University Press, pp. 142–166.

Saker, Michael and Frith, Jordan (2022), 'Contiguous identities: The virtual self in the supposed metaverse', *First Monday*, 23 February, https://firstmonday.org/ojs/index.php/fm/article/download/12471/10612. Accessed 27 March 2022.

Tam, Chul-Ling (2018), "Why we should stop labelling people climate change deniers", *The Conversation*, 3 December, https://theconversation.com/why-we-should-stop-labelling-people-climate-change-deniers-105555. Accessed 27 March 2022.

Thomas, Amanda, Cretney, Raven and Hayword, Bronwyn (2019), 'Student strike 4 climate: Justice, emergency and citizenship', *New Zeland Geographer*, 75:2, pp. 96–100.

Watts, Jonathan (2019), 'Greta Thunberg, schoolgirl climate change warrior: 'Some people can let things go. I can't'', *The Guardian*, 11 March, https://www.theguardian.com/world/2019/mar/11/greta-thunberg-schoolgirl-climate-change-warrior-some-people-can-let-things-go-i-cant. Accessed 23 March 2022.

Willson, Michele (1997), 'Community in the abstract: A political and ethical dilemma?', in David Holmes (ed), *Virtual Politics: Identity and Community in Cyberspace*, London: Sage, pp. 145–162.

CHAPTER 4

The Medium is the Environment—COP26 Memes as Hypocritical Resistance

Adalberto Fernandes

Introduction

Mass media, social media, and new media are decisive for the management of human populations' expectations and actions regarding environmental issues because they frame what and where is the danger (Demos 2017). For example, the communication of short or long-term fatal impacts on the environment leads government and people to respond with more or less urgency and invest more or fewer resources to react to those impacts. Contrary to the immediate environmental crisis, like volcanos and storms affecting nearby populations, the communication of the risks related to climate change seems to have a temporal and spatial scale where the human intervention seems less urgent. Given its impacts unfold over a lengthy period, and to counteract those impacts it is necessary for a global effort that mobilizes collectives, governments, institutions, international

A. Fernandes (✉)
Centre for Organisational and Social Studies (CEOS.PP), Porto Accounting and Business School—IPP, Porto, Portugal
e-mail: adalberto.castro.fernandes@gmail.com

D. Binns and R. Najdowski (eds.), *Confronting the Climate Crisis*, Palgrave Studies in Media and Environmental Communication,
https://doi.org/10.1007/978-3-031-89606-4_4

organizations and global corporations, we can say the environmental crisis has a weak rhetorical power. What we mean is it is difficult to persuade someone to act on matters where the scale of intervention is massive because actions will seem to have a low impact or an impact that is not immediate and, therefore, cannot be readily testified. The importance of communication lies in its capacity to provide 'commensurable' mediations of varied scales of time and space. This thesis is almost a given in communication studies: for instance, mass media amalgamated distant spaces by disseminating information from distance sources in shorter periods of time (Lippmann 1922); the written word allowed the discourse to survive the distance and the death of the author (Derrida 2016). Communication is, thus, one important topic to think about how the diverse time and space scales of the environmental crisis are translated.

We must focus on the ways the environment and climate are being communicated in the different media. Depictions of climate crisis—as with almost any image—are produced, framed, and interpreted with certain social, cultural, and political objectives, and cannot be taken unproblematically as 'pure representations' of nature (Carruth and Marzec 2014). This means a certain 'human hand' is already intervening in the representation of 'nature'. However, to separate humans and nature, divided by the mechanism of representation—where nature is equated with presence and humans represent it—, is a division that should be addressed critically. Recent work in the humanities tries to re-think the natural science concept of 'Anthropocene' (Crutzen 2006), which brings to the fore the impact human actions (anthropo-logical actions) have on the environment and how humans suffer from that impact. Discussions about the anthropocene topic, as it is approached in the humanities by Haraway (2015) or Latour (2014), puts in question a clear-cut division between humans and non-humans, between humans and nature. Anthropocene is, then, too a 'Naturalcene' or 'Nonanthropocene', provided we cannot think humanity as wholly detached from non-humanity or nature.

Questioning the human/non-human division can be operationalized if we try to understand how the environment communicates itself *through*, instead of being communicated *by*, media, social media, and new media. If the materiality of media—the smartphone or the human body that uses it—is a composition of human and non-human elements (because the smartphones have minerals and human bodies have bacteria), this means environmental communication is not to be considered solely in 'representational' terms. Media is already a part of nature, the part of nature that communicates. To communicate the environment is not to represent something standing "out there" in a pristine form, waiting to be captured and presented again (represented). The environment becomes, on the contrary, something always distinct when different media are processing it because it is the environment changing into diverse communicational forms. This means, given the addresser, addressee and means of communication are the same—i.e. the (media) environment—, to produce a message, to change its content and form is also to change the environment that makes the message possible. Given communication is made of environmental materials, which implies communication is (in) the environment and changes it, this means environmental communication is the environment 'talking' with itself. Our position is in line with the work of Peters (2015, p. 3), where the "old idea that media are environments can be flipped: environments are also media". The sense in which natural elements mediate things can be traced in the history of ideas to Newton, especially when his work was integrated into biology by Lamarck, to explain the role of the natural medium ether in the action at a distance between different physical individuals, giving ether was used to explain where light travelled to affect physiologically the eye that perceives it, because ether was considered being present in the eye and in the air (Canguilhem 1952). The environmental problems cannot be fully tackled and understood without an approach to media as a "media ecology" (Scolari 2012). Media are part of the environment. They not just the expressive, symbolic, or representative part of an *uncommunicative* environment (Wickberg and Gärdebo 2022). Otherwise, we could not manage and comprehend the environment if communication wasn't already produced by the environment. Bees communicate between themselves to build and react to the environment, but also signal us that something is wrong with the environment. An environment that does not communicate cannot tell us what is urgent and what needs to be done if we want to survive. In the same way, media which is not part of the environment, made of its resources, does not exist. Media is from this world, not a transcendent moment without its own carbon footprint.

Materials and Methods

Although we have an important amount of academic literature about environmental and climate topics in the media, there are still a small fraction of studies dedicated to the relation between the Anthropocene and the media (Sklair 2020). We want to contribute to this debate about the environment as media, against a "representative" logic (for non-representational theory, see: Vattimo 2011; Derrida 2016; Foucault 2005; Rorty 1979), and stark divisions between humans and nature. We will do this through a case study based on a specific medium: the 'meme'. We will analyse memes criticizing the 2021 United Nations Climate Change Conference—COP26 to know how they may constitute a source of a resistance discourse. The critique against the politics of COP26 is exercised through memes, a type of medium that is fast, international (given its prolific use of a globally shared visual culture, and the English language), and works through disruptive re-appropriations of mass media messages and images. This case study will allow us to answer the following question: *If memes are a type of communication produced by the environment, given communication is an assemblage of human and non-human proprieties, what type of (communicational) environment are they producing?*

The meme constitutes a type of communication that can be understood as behaving like natural organisms (Dawkins 1976) given the way they produce 'genealogical lineages' of images that can be traced by observing the ways the images mutate. More than a biological metaphor, granted the environment already behaves communicationally, we can say memes are a 'wild' communicational "species" in a communicational ecosystem made by the internet. Memes can be considered being wilder than the 'more domesticated' species of the media ecology like traditional media (TV, radio, press), or ritualized communication (court, church, military, etc.), given they can be produced by anyone and are not restrained by formal and content constrains. Memes are used as a humourous way to approach environmental issues (Kaltenbacher and Drews 2020; Davis et al. 2016). They also play with the scales of time and space by being a fast Do-It-Yourself (DIY) hyper-inter-text, creating large communities around simple compositions of image and/or text that can be mobilized for environmental issues (Zhang and Pinto 2021; Ross and Rivers 2019). This speed is crucial for the communication of the urgency of the environmental crisis.

We will qualitatively analyse three memes about COP26 published on the site "9gag.com", one of the most viewed and used website for the publication of memes (Lo 2018). The qualitative method is appropriate given a quantitative analysis of the images cannot capture the dense web of meanings memes have to offer. Memes are made of an intrinsic ambiguity, the result of an absence of redundancy and/or rich interpretative clues. This allows memes to be shared widely because they do not demand a lot of background information to be understood, but they also produce a loss of meaning. A quantitative analysis of memes is useful to calculate the number of ways in which images are arranged and composed and to identify the dominating topics. However, only a qualitative analysis allows us to understand how the meme mediates specifically the environment by analysing how the meanings are constructed by a heterogeneous composition of images and text, because this heterogeneity resists straight and univocal categorization needed to operationalize quantitative analyses.

The theoretical assumption guiding our qualitative analysis, discussed in the introduction, is that memes are a species of the environment. They are a mode of communication *of* the environment and not just *about* the environment. We are methodologically "suspending", instead of completely abolishing, strict divisions between human and non-human,—following Actor-Network Theory (Latour 1996)—, suspending, thus, absolute divisions between memes and the environment. The aim is to understand memes not as a "representation" of nature, but as a kind of "nature", something which is part of the environment. This methodological choice helps us avoid defending that memes are not part of this world, or the claim environment does not include human artefacts, like memes. It also helps us avoid putting communication in a transcendent position in relation to the environment, allowing us to see how it affects and is affected by the environment, which is also communicative. This methodological suspension enables us to perceive how memes operate on the environment instead of merely representing it. For the formalist description of the memes, we will briefly introduce them by focusing on its "informational value", "salience" and "framing" dimensions (Kress and Leeuwen 2020). In other words, form determine what memes are about, what the elements dominating the image are, and how the placement of heterogeneous elements establishes or breaks frontiers between them.

Hypocrisy travels fast

One of the dominant topics of the posts and memes about COP26 is "hypocrisy", understood as the breakdown of the rhetorical dimension of *ethos*, of character, which results from contradictory actions that jeopardize the coherence of the speaker's ethical position. Many news pieces shared on the 9gag website highlight the fact the world leaders travelled to COP26 by plane, considered a high pollutant means of transportation to an event which has the aim of fighting pollution. In the meme intituled "How dare you" (Fig. 4.1), the critique of this hypocritical political position is made with disparate elements: a screenshot from the *film Mad Max: Fury Road*, flags from different countries, the logo of COP26, and a picture of Greta Thunberg's head. COP26 logo contextualizes the critique's target, namely, the reason why all the high pollutant cars of *Mad Max Fury Road* are moving somewhere. Participating countries' flags are juxtaposed with the various polluting vehicles in the image. It is relevant to notice the USA flag placement in the largest vehicle, which is also an actor with a 'flamethrower-guitar' in his hands, suggesting the USA are a central target of the critique.

The 'framing' of the composition unites distinct worlds of meaning by, paradoxically, an addition of frames: the head of Thunberg, the logo of COP26 and the flags are not integrated into the picture, they are juxtaposed in a way that makes visible what is and what isn't part of the movie's screenshot. These 'frontiers' between images characterize the rapid, do-it-yourself (DIY), and amateur style of memes, and they do not entail necessarily a separation between elements, otherwise, the images would not be intelligible in the context of COP26, the pollution of world leaders and the activism of Thunberg. Interestingly, the movie *Mad Max: Fury Road* is a post-apocalyptical action film about a world becoming a desert wasteland following a societal collapse from warfare over resources, and the desire to escape to the "Green Place", an idyllic land. However, the movie is a celebration of cars, with their DIY 'steampunk' allure of the

Fig. 4.1 Deerdayvile, *How dare you*, 30 December 2021, Montage image/meme. Copyright-free. http://9gag.com/gag/a81850O

mechanical bricolage of motor vehicles. Vehicles are an attractive aesthetical component of the movie but, in the meme, they are used as the worst type of transportation that nations could use to go to COP26. The meme itself produces this transfiguration of COP26 and the movie, showing the typical heterogeneous collage of elements this medium allows, and making a unique meaning not previously present in the isolated elements of the image. In this sense the meme is not representing something that is "out there", this composition doesn't have a counterpart, a referent outside of the image, even if its isolated elements (the movie, cop26, flags and Thunberg) exist outside this meme.

Facing against the cars we have a picture of the head of an angry Greta Thunberg, going a way to explain the title of the meme—"How dare you"—a visual and discursive reference to Thunberg's speech at the 2019 UN climate action summit in New York. This speech is known for the aggressive style that characterized the discourse of Thunberg, and the angry facial expression is mobilized by the image. The placement of Thunberg's head against the vehicles reinforces the critical stance of the meme. The visual disproportion of Thunberg's head and the size of the cars put Thunberg, at the same time, inside and outside the film's screenshot. The activist is not in this post-apocalyptic landscape, but this odd composition is commonplace in memes, which tend valuing placement above framing. This means the proximity of visual elements, without concern with creating the 'right' proportions or framing all the elements coherently, is sufficient for the mode of mediation of meme characterized by a 'displaced placement'. Thunberg is not supposed to be there. It is also obvious she is not part of the movie. The visual disproportions prove she is not there. She is not 'originally' part of the image. However, the mode of mediation of the meme tells us the formal aspects of visual placement don't prevent producing a new world of meaning. A world where the head of Thunberg may make sense against motor vehicles in the desert with 'untidily' pasted flags. In this world of meaning, flags are a pollutant entity, flags have a driving-agency, and they must face a disproportionate activist's head.

The fictional world of *Mad Max: Fury Road* receives another layer of 'quasi-fiction' or 'semi-fiction', which comes from the mixture of a non-fictional world of COP26 and the activism of Thunberg. The clash or encounter between fiction and non-fiction displays an original tension, a new image, where non-fiction (flags, COP26, Thunberg) and fiction (the movie) are assembled to create an image which it is not just about the movie and not just about the environment. It is a "serious fiction" a "real fiction", where the fictional world acquires seriousness by being used to criticize environmental politics, and the political critique receives the aesthetical force of the fictional world of films. This encounter produces implausibility, the image is unbelievable, which contributes to generate a serious critique showing that the real situation, the non-fictional world, is already unbelievable, hypocritical (e.g.: world leaders flying to COP26), and only fiction can help us make the unbelievable believable. The meme mobilizes a "hyper-critical" strategy to fight "hypo-critical" environment politics. Without the fictional visual resources that hyper-aggravate the consequences of present hypocritical politics, that is, a critique reduced to non-fictional tools of meaning production, the meme might not be able respond to the politics of the environment which have become (hyper-)unbelievable and gained, thus, a fictional aspect that makes it harder to believe they are actually happening. In other words, to oppose the fictional appearance of current hypocritical environment politics, the meme (hyper-) critique engages with fiction to fight with the same power of (hyper-)un-believability.

Despite the hypercritical strategy of the meme, the political hypocrisy of COP26 is answered with a hypocritical image. The meme tries to move between fiction and non-fiction to fight the non-fictional world while producing a fictional composition. The hypocritical nature of the meme, in terms of its fictional/non-fictional status, results from the hypercritical strategy, a strategy defined by the visual hyper-exacerbation of the state of affairs of environmental politics. The meme composes an image where politics is already post-apocalyptic, but only fiction can give that possibility of composition with its visual culture associated with this type of film genre. *Mad Max: Fury Road* is an apocalyptic fiction pointing towards the failure of environmental politics, where the natural resources are exhausted, and the anarchy of the vehicles in the desert signal the loss of a political project that is not just one of survival. Memes gives visual materiality, from its "semi-fictional and semi-serious" place of image production, to the impacts of world leadership in environmental politics, but this is an image that can only be produced by a non-apocalyptic mode of visual production. The film industry and the meme community produce fiction because their non-fictional conditions of production are not at stake. You can only paint the destruction if your canvas is intact. The hypercritical strategy of the meme, possible by its fast and almost imperceptible movement between fiction and non-fiction, is hypocritical given the apocalypse has not arrived yet (contrary to what the meme tries to make us believe) because it cannot arrive when it is still possible producing images of it. Although there are plenty of images about the apocalypse, it can also be said that *when the apocalypse unfolds there would be no images of it*, in the sense there would be no conditions to depict it once we are occupied with surviving, that is, handling the brutal images the apocalypse is giving us. The apocalyptic narrative genre is an assemblage of non-apocalyptic conditions (e.g.: conditions for actors to survive in the desert), and the meme uses those images to develop a hyper-critique of the apocalyptical world leaders' behaviour. However, the meme is a hypocritical critique because those images result from the non-apocalyptical film industry. This apocalyptical meme gets its hypercritical power from non-apocalyptical means of movie production. The meme is also hypocritical because, through a heavy use of fiction, fails addressing the ongoing state of affairs, missing the object of critique (e.g.: world leaders are not promoting the political post-apocalyptic anarchy. They do not need that to destroy the planet).

In a more general sense, the meme, as a medium, not only represents the apocalypse, but it is already apocalyptical. When we do not understand a meme (and it will be explained above why that is likely to happen) this shows that it already operated a 'micro-apocalypse' in the ecology of media. This is not exclusive of memes. Our incapacity to read Latin, or use MS-DOS, already signals the destruction of the media environment. However, what is distinctive in the 'meme micro-apocalypse' is the speed and vulgarized way in which the destruction occurs, given more people can produce and publish 'fictional-non-fictional' visual compositions. Memes are increasingly perishable and irrelevant given they demand an impossible interpretative community that, to survive as such, should be able to maintain a shared and encyclopaedic visual culture that could be updated with the same pace of meme production. The attempts of producing apocalyptic images to respond to the hypocrisy of environmental politics have given place to an accelerated media apocalypse marked by an incapacity to understand images, leading to an increased rate of visual culture destruction, and giving way to a 'visual mass extinction', which is rapidly produced in the world of memes. To attack political hypocrisy it is necessary for the environment, in the way it mediates itself (considering memes as a species of the media ecosystem), to produce visual extinction, which is the extinction of environmental resources because images are part of the environment.

Circular economy of media car-culture

Another reference to the world of cars is made in the meme "COP26 summit be like" (Fig. 4.2), which uses up a "meme template". That is, the use of a familiar image or combination of images and/or text gives a common form to new events. In the same way a newspaper has a title in large letters and a lead with the "who, when, where, what, how and why" of the news, the meme also has, with multimodal elements, its structured ways of receiving varied information. It is the repetition of a specific multimodal format to couch and interpret different content which characterizes a meme as a meme. In this meme, we have another Anglophone pop culture reference, in this case, the automotive lifestyle/entertainment show *Top Gear* (2002-present). This show about cars has become 'meme-ified' in the following way: co-host James May claimed the manufacture of a certain brand of car was delayed. Co-host Jeremy Clarkson reacted with an ironic "oh no", shifting suddenly to another topic, saying "anyway" to indicate how little interest he had in the mentioned car. In the context of COP26, this meme has been used in the following mode: two images of Clarkson with the words "oh no" and "anyway", are presented with the new information that "The earth temp is rising, sea levels are rising, greenhouse gases are rising—Oh no!" and with the response "Presidents with their private jets—Anyway".

Similarly to the *Mad Max: Fury Road* movie, the *Top Gear* image is also linked to the world of cars, which is not primarily concerned with the environment but with the aesthetics and market of motor vehicles, something that does not prevent the image from being re-appropriated for proposes of critique. *Top Gear* is a type of show that stimulates the purchase and admiration of cars, making it part of the fossil-capitalist political economy of media. However, the dialogic interaction of the hosts offered a linguistic and visual template the meme community uses for divergent objectives, regardless of the original context of the show. If we approach meme as a medium which is a species in the ecology of media, its 'nature' is one that transforms pollution into a critique against pollution. Besides using the energy and carbon footprint produced by digital devices and the internet, necessary to produce memes, the meme re-uses the car-culture to oppose the environmental problems associated with that

Fig. 4.2 Showkey, *COP26 summit be like*, 3 November 2021, Montage image/meme. Copyright-free. https://9gag.com/gag/a21NnEY

culture, adopting the power the aesthetics of post-apocalyptical cars and the notoriety of automobile TV Shows to fight the environment politics identified with that culture. This suggests the communication struggle for the environment is not a clear separation between the polluting media and the non-polluting media, but the production of a hybrid that uses the power gathered by polluting forces and the aesthetics and political economy of media that is associated with the car industry, to re-create its own force of resistance with a circular media economy. The meme is a type of mediation that not only circulates—given its constant re-utilization of images as "meme templates"—, but can also develop a circular ecology, by re-using media linked to the pollutant car-culture to resist pollutant culture promoted by the world leaders travelling to COP26. Our analysis of these memes is another strange species of the media environment, which points towards not only to its intertextuality but also to its 'inter-media-ecology', by producing 'media circularity', a 'circular economy of media'. Our analysis also promotes the pollution associated with the construction of these images, which could be reduced if we did not use the images to illustrate our point. However, we would lose the persuasive capacity to convince the reader, a capacity performed with the assistance of the hegemonic presence of these images that allows convincing more people. Our text is not a representation of memes, but it is a re-circulation of the media circularity produced by memes. It is part of the (media) ecology by recycling and polluting to persuade.

The Anti-Transitioning Politics Paradox

The presence of pop culture references from the blockbuster movies is also explicit in another meme that targets the hypocrisy of COP26 with the following title: "Jets, luxury cars, fining dining. Lecturing us to save the planet" (Fig. 4.3). The title already introduces us to a discourse where contractions or political incoherence is brought to the fore. A reference to a news piece of *The New York Times* is placed at the top of the meme, giving it a high salience, which is reinforced by the fact the newspaper is an informational source, giving credibility or veracity to the meme. According to the news: "The climate summit's own carbon footprint will be substantial, a report says". The news, which works as a producer of credibility, is reinforced by a mention of a "report" which confers another layer of informational value to the meme. A popular culture element of the meme is "Yoda", a fictional character from the *Star Wars* movies. Yoda is accompanied by a transformed phrase, "the hypocrisy is strong with this one", which re-appropriates the original popular meaning: "the force is strong with this one". This movie character is associated with sapience, knowledge, which contributes to the information value of the meme because what is said by Yoda has the mark of intelligibility, given it comes from a character with wisdom. In this visual and textual fictional/non-fictional world, where Yoda reacts to an American newspaper, the meme creates a "semi-fiction", "quasi-fiction" or "serious-real fiction".

Yoda's meme indicates a clear tension and uneasiness with the need of increasing the carbon footprint while fighting pollution. COP26 should be environmentally pure. Resistance against climate change cannot produce pollution. The meme suggests a world where fighting for climate change will not produce pollution, as if the politics of climate were clearly divided between pollutants and non-pollutants. This suggests the impossibility of a "transitioning politics", where pollution must stop immediately, and no pollution is tolerated even to fight pollution. The title, "Jets, luxury cars, fining dining. Lecturing us to save the planet", points not simply to an anti-transitioning politics but also a class-driven struggle, where the world leaders are depicted as a privileged group of people lecturing the non-privileged to take care of the environment. The rich don't have the "economic ethos" or economic legitimacy to lead the masses to save the planet, given the unjustified inequalities and power differentials between people that must obey and those that can command. Not only there is a refusal of transitioning politics, but also a critique

Fig. 4.3 Rohittaurus, *Jets, luxury cars, fining dining. Lecturing us to save planet*, 15 November 2021, Montage image/meme. Copyright-free. https://9gag.com/gag/aZrxwrz

of ecological politics based on economic inequality. This is an interesting perspective, given *Star Wars* is made by a film and entertainment industry which has generated immense revenues from its franchising and sequels, producing pollution and inequalities in this sector. COP26 memes, thus, not merely re-appropriate and reinvent the media supporting and stimulating the automobile industry (*Mad Max: Fury Road* and *Top Gear*), but they also intervene in the media ecology by re-using images from a film industry which yields economic inequalities to criticize economic injustice.

Once again, the meme rehabilitates the material and economic resources used to produce inequalities and pollution in the world to oppose the unjust and dangerous state of affairs. The media ecosystem not only produces pollution, through the use of cars in movies and in TV shows promoting the car-culture, or leads to inequality by concentrating the economic resources to produce blockbuster films. Through memes, "re-cycled" media puts in question that same polluting media ecosystem. Memes also leave their carbon footprint, but they employ it to question the carbon footprint generated by the car-culture and capitalist film industry. Resistance is done with the polluting images, and not with a production of totally new, non-pollutant media because the pollutant media is the one that reaches more people by being hegemonic in our culture. The memes need to use pollution if they want to influence the politics of the environment massively.

Opposition to a "transitioning politics" through memes will be contradictory because the meme is, in itself, a "transitioning (media) politics". It uses pollutant media to combat pollution. A meme that refuses to use the high shared visual culture of polluting media reduces drastically its chances of producing an impact on the polluting media. The impact of certain media has created not solely the conditions for the destruction of the environment (*Mad Max: Fury Road* and its car culture) but also the conditions for the critique of its pollutant behaviour (the 'meme-ification' of *Mad Max: Fury Road against car-culture*). The mode of communication of memes is one of "circularity", stimulating a "circular economy of images", given they fight the pollutant images by re-appropriating those images against them. This circulation is, thus, always a politics of transition. To reach large audiences is necessary to use the pollution produced by media (movie and TV industry) because this pollution allowed to arrive at broad masses, and the meme-ification of those images contribute to propagate those same pollutant media, while fighting against the current hypocritical environment politics of COP26.

Media as another name for the environment

What does this mean in terms of a 'strong' reading of media ecology? A reading which does not consider the meme as a transparent representation of an external world of reference, as if the world was unaffected by the way it is mediated, or as if media had a 'weak' link with the environment. In other words, the "easy problem" of the relation between media and environment is how media representations of the environment give accurate or inaccurate descriptions of an external environment. This presupposes media and environment are external to each other and a relationship of correspondence, or adequacy, is possible. On the contrary, the "hard problem" of media considered as a part of the environment is: the meme, when it criticizes the hypocrisy of the politics of the environment, gives us an environment which is itself hypocritical. The genre "environmental critique meme", which is a species of the (media) environment, shows that the environment utilizes pollution as a response to dangerous environmental politics. This claim is justified if we accept memes are part of the environment, with material impacts on it, given their existence is not inconsequential to the environment. A meme is solely possible with environmental resources and affects those same resources: the electricity of the computer, the brain cells of the user, the plastic of the laptop, the finite physiological process of attention to persuade people to act in a certain way in the environment.

The study of media, and the particular study of memes, shows that the environment—if taken as including its mediation dimension—is ambiguous; mobilizes fiction and non-fiction to manage the political hypocrisy of COP26; does not give us a 'natural truth' to avoid hypocrisy; and uses hypocrisy as a natural survival resource for humans and non-humans. Memes indicate that the environment fights *political hypocrisy* with *fictional hypocrisy*. *Hypocritical* in the sense the meme does not seem to recognize its own pollution and mobilization of powerful images originally linked to a polluting culture, and *fictional* because the meme proposes a world of meaning where fiction and non-fiction are mobilized to forge a hyper-critique matching the hyper-unbelievable non-fictional state of affairs (e.g.: the destruction of the environment, or the behaviour of world leaders). This hypocrisy and fiction are not just an exclusive characteristic of memes, but they are part of the environment, even before memes have appeared as a media species. We do not need politics or memes to know the environment is hypocritical and fictional in its communicative mode. For instance, orcas deceive seagulls by throwing fish to attract and eat them. In analogous terms, the meme functions like this fish. This fish is not just simply food for seagulls, given the offer of fish is a *fictional* and *hypocritical* gift to attract them. Not only is not a real offer, but also entails death instead of nutrition. However, the fish does not cease to be food, *non-fiction*, otherwise it would not work as a bait. Likewise, the meme is not simply a fiction, it has non-fictional effects obtained through fictional means. Media and fiction are not representations and separate entities of a non-mediated or non-fictional environment. They are part of it. We can produce non-fiction about fiction (e.g.: literary critique, or sociology of art—two species of the academic media environment). Art makes us cry real tears, we pay real money to see fictional works, and fiction takes time and occupies space. In sum, fiction is part of the environment. Memes' hypercritical strategy based on fiction is only powerful if mobilizes persuasive polluting images; they resist the apocalypse by being an apocalyptical medium itself, using and destroying visual environment resources at a rapid pace; and they want to produce non-fictional effects on the world through fiction to match the fact non-fictional environment politics have become so unbelievable (just like fiction). The same could be said of the present chapter, understood as part of the (media) environment, which is hypocritically polluting by using pollutant media to achieve a hypercritical strategy, and pretending to move between fiction (the memes analysed) and non-fiction (the analysis of meme) to obtain a non-fictional believable effect on a world whose massive destruction has become unbelievable.

Fiction is one important mechanism of the circular economy of the media environment. Fiction allows a re-utilization of images displaying hypocritical and hypercritical strategies, which are the ways the environment responds to its own communicative behaviour. It is because the environment "talks" about itself, communicates itself, instead of its communication being something out of this world, that the environment needs fictional and hypocritical strategies to solve its communicative crisis, given it is (hyper-) criticizing itself once we approach memes and politics as environmental species, not exterior elements to the environment (otherwise how could they produce any influential difference in it?). Fiction allows the environment to escape gross communicative contradictions that could contribute to its destruction. Fiction transforms media capitalism, car-culture and the pollution of digital communication, through the mediation power of the meme species, into a circular economy of communication to fight its hypocritical crisis that can lead to its destruction. Given fiction is from this world, part of the environment,—because there are material conditions and effects linked to fiction—, to combat political hypocrisy with fiction is to make another hypocrisy, as if fiction was a transcendent representation of the world without its own carbon footprint. Why? Because memes use pollutant fictional images to criticize pollution in a non-problematized way, as if that use was without consequences for the environment.

The meme hypocrisy results from a hypercritical strategy that seeks to aggravate, with apocalyptical fiction, the potential consequences of the hypocritical politics of COP26, mobilizing apocalyptic images originally designed by the hegemonic media industry that contribute to the car-culture and inequality in the world. However, this hypocrisy puts in motion a circular economy. The environment employs fiction to create a real circular economy of communication, where fiction (*Mad Max: Fury Road*, *Star Wars*) is mobilized to respond to non-fictional threats of the environment politics hypocrisy (world leaders flying to COP26) and give hypercritical semi-fiction or serious-fiction (the meme) the communicative power (formed by the combination of fiction and non-fiction) that the high pollutant media entertainment ecosystem has, which allows it to reach large audiences. Fiction is a communicative resource of the environment to "get out" of the environment without actually escaping it. Fiction produces a sensation of vertigo by changing our perspective even if do not move an inch. The head of Thunberg is not in the desert of *Mad Max: Fury Road*, but by being placed in a meme, this new "quasi-fiction" or "serious-fiction" made of a mixture of fiction and non-fiction, allows the environment, in its "meme-ified behaviour", to communicate a hypocritical image (i.e. the activism join forces with the pollutant and capitalist images of the film industry) that allows the environment to produce a circular media economy, which mobilizes communicative pollution to hypercriticize that same pollution.

The high risk of not understanding a meme (not having the visual repertoire to identify a screenshot from *Mad Max Fury Road*, the host of *Top Gear*, or *Star Wars*), thus leading to the extinction of visual culture (with impacts on the environment once we forget the car-culture and capitalist political economy of media associated to that visual culture), conducting us to an apocalypse of images, is produced singularly by this medium. This visual world is composed of a fast composition of images, whose speed justifies why visual elements don't fit "tightly", given the untidiness characterizes the type of mediation of memes. There is no time and no need to devote large amounts of it to create a perfected composition. Meme's meaning is obtained by the simple proximity of visual and/or textual elements in the same image. Visualizing a meme, as in any case of reception, is characterized by the active work of filling the meaning gaps such placement of disparate visual elements produces. This interpretative filling work of the receiver is aided by the agenda of COP26. However, COP26 provides a context that does not decide the full interpretation of the meme, given the meme is characterized by the "loss of meaning". The meme, while being a highly socially shareable medium by travelling with little information as possible, has a power of dissemination based on not explaining everything. It is based on a lack of redundancy and explicit interpretative clues. This makes the meme, in informational terms, "lighter" by reducing the amount of information to decode. It is unnecessary to recognize the screenshot from the picture *Mad Max: Fury Road*, the car show TV host, or if Yoda is a figure of wisdom, to understand we are in the presence of a critique of COP26.

The meme also reduces the "decoding" (Hall 2007) tools available by the image. We can thus lose sight of the automobile-culture and capitalist political economy of the film industry if we do not have enough visual culture to identify the images of pop culture that contribute to the destruction of the ecosystem, a destruction that makes those images possible, leaving us without tools to understand how meme takes part in this destruction and, at the same time, resists it. The circularity promoted by the meme produces its own destruction of the (media) environment, that is, an extinction of the visual culture associated with pollutant media. We run, thus, the risk of putting in circulation memes that hypercriticize the environment politics hypocrisy with pollutant media, leaving us "blind" to the pollution made by our own hypocritical critique. This extinction of visual culture shows that if we do not approach memes as a species of a (media) ecology, as if they were just an external representation of the environment, we lose sight of the impacts memes have on the pollutant mode of production of images (recycling the car-culture and the big film industry) and how those pollutant images make the meme possible (making the mode of critique of the meme hypocritical).

Concluding remarks

Memes create new ways of environmental transformation. They show how the environment communicates itself with pollutant media, hypocrisy, and capitalism, to resist the dangerous environment politics. Without an analysis of the environmental behaviour of memes, we lose an important species of the (media) ecosystem that can help us understand how to do environmental communication that forces us to be critical about absolute separations between environment and communication, human and nonhuman, fiction and non-fiction, hypocritical and hypercritical, pollutant and non-pollutant.

Acknowledgements This work is financed by Portuguese national funds through FCT - Fundação para a Ciência e Tecnologia, under the project UIDB/05422/2020 and FCT-TENURE — 1st Edition - Ref. 2023.11412.TENURE.001

References

Canguilhem, G. (1952) *La Connaissance de la Vie*. Paris: Librairie Hachette.

Carruth, A. and Marzec, R.P. (2014) 'Environmental Visualization in the Anthropocene: Technologies, Aesthetics, Ethics', *Public Culture*, 26(2 (73)), pp. 205–211. Available at: https://doi.org/10.1215/08992363-2392030.

Crutzen, P.J. (2006) 'The "Anthropocene"', in E. Ehlers and T. Krafft (eds) *Earth System Science in the Anthropocene*. Berlin, Heidelberg: Springer, pp. 13–18. Available at: https://doi.org/10.1007/3-540-26590-2_3.

Davis, C.B., Glantz, M. and Novak, D.R. (2016) '"You Can't Run Your SUV on Cute. Let's Go!": Internet Memes as Delegitimizing Discourse', *Environmental Communication*, 10(1), pp. 62–83.

Dawkins, R. (1976) *The Selfish Gene*. Reprint edition. Oxford University Press.

Demos, T.J. (2017) *Against the Anthropocene: Visual Culture and Environment Today*. Berlin: Sternberg Press.

Derrida, J. (2016) *Of Grammatology*. Baltimore, Maryland: John Hopkins University Press.

Foucault, M. (2005) *The Order of Things*. New York: Routledge.

Hall, S. (2007) 'Encoding and decoding in the television discourse', in A. Gray et al. (eds) *CCCS Selected Working Papers—Volume 2*. London: Routledge.

Haraway, D. (2015) 'Anthropocene, Capitalocene, Plantationocene, Chthulucene: Making Kin', *Environmental Humanities*, 6(1), pp. 159–165. Available at: https://doi.org/10.1215/22011919-3615934.

Kaltenbacher, M. and Drews, S. (2020) 'An inconvenient joke? A review of humor in climate change communication', *Environmental communication*, 14(6), pp. 717–729.

Kress, G. and Leeuwen, T. van (2020) *Reading Images: The Grammar of Visual Design*. 3rd edn. London: Routledge. Available at: https://doi.org/10.4324/9781003099857.

Latour, B. (1996) 'On actor-network theory: A few clarifications', *Soziale Welt*, 47(4), pp. 369–381.

Latour, B. (2014) 'Agency at the Time of the Anthropocene', *New Literary History*, 45(1), pp. 1–18. Available at: https://doi.org/10.1353/nlh.2014.0003.

Lippmann, W. (1922) *Public Opinion*. New York: Harcourt, Brace.

Lo, A. (2018) *9GAG: From side project to internet phenomenon*, *CNNMoney*. Available at: https://money.cnn.com/2018/04/26/smallbusiness/9gag-ray-chan/index.html (Accessed: 7 April 2022).

Peters, J.D. (2015) *The Marvelous Clouds: Toward a Philosophy of Elemental Media*. Chicago ; London: University of Chicago Press.

Rorty, R. (1979) *Philosophy and the Mirror of Nature*. Princeton: Princeton University Press.

Ross, A.S. and Rivers, D.J. (2019) 'Internet memes, media frames, and the conflicting logics of climate change discourse', *Environmental communication*, 13(7), pp. 975–994.

Scolari, C.A. (2012) 'Media Ecology: Exploring the Metaphor to Expand the Theory', *Communication Theory*, 22(2), pp. 204–225. Available at: https://doi.org/10.1111/j.1468-2885.2012.01404.x.

Sklair, L. (ed.) (2020) *The Anthropocene in Global Media*. 1st edition. Abingdon New York, NY: Routledge.

Vattimo, G. (2011) *A Farewell to Truth*. New York: Columbia University Press.

Wickberg, A. and Gärdebo, J. (eds) (2022) *Environing Media*. 1st edition. S.l.: Routledge.

Zhang, B. and Pinto, J. (2021) 'Changing the world one meme at a time: The effects of climate change memes on civic engagement intentions', *Environmental Communication*, 15(6), pp. 749–764.

CHAPTER 5

Engendering Care for the Environment Through Podcasts

Lili Pâquet and Wyatt Moss-Wellington

Podcasts, an amalgamation of 'iPod' and 'broadcast', were first developed in 2004 and have increased in popularity in recent years. The accessibility of podcasts, often streamed using an app on a listener's smartphone so that it becomes 'a personalised radio [...] in your pocket' (Riordan 2018: 68), allows audiences to 'time-shift and place-shift content, meaning a user can tune in to the podcasts they choose, whenever, and wherever she wants' (Markman and Sawyer 2014: 20; McClung and Johnson 2010: 83–91).

A subset of podcasting investigates environmental change. In this chapter, we analyse and compare episodes of the environmental podcasts *Outrage + Optimism* and *Sustainababble* produced in 2020–21. We ask how these podcasts present an emotionally resonant narrative to provoke heightened care for, and delight in, the environment, and how those

L. Pâquet · W. Moss-Wellington (✉)
Department of Creative Arts and Communication, University of New England, Armidale, Australia
e-mail: Wyatt.Moss-Wellington@une.edu.au

D. Binns and R. Najdowski (eds.), *Confronting the Climate Crisis*, Palgrave Studies in Media and Environmental Communication,
https://doi.org/10.1007/978-3-031-89606-4_5

emotions are related to the more negative affect of activist rhetorics (for instance, a motivating anger, guilt, or shame). Through analysis of the chosen podcasts using the prisms of rhetorical identification and political emotions, we show how global, diverse audiences may identify with environmental arguments, find them emotionally resonant and become empowered to make changes to the world. We find that environmental podcasts aim to strike a balance between anger, hope and humour, using emotional, aural narratives to build a feeling of community, through which listeners are given motivation for personalized environmental adaptation and activism.

This personalization of activist rhetoric is key to understanding the intended influence of environmental podcasts. People of differing professions and life circumstances hold different responsibilities for climate action. Environmental podcasts thus seek to motivate individuals to pursue climate action in their own worlds, rather than prescribing one manner of best-practice activism. The podcasters instead adapt their messages through the capabilities of the aural medium to engender an activistic disposition; we reveal the mix of emotions environmental podcasters incite to contribute to such a disposition.

Methodology

Our analysis focuses on two representative environmental podcasts, *Outrage + Optimism* and *Sustainababble*. *Outrage + Optimism* is hosted by Christiana Figueres (Executive Secretary of the United Nations Framework Convention on Climate Change 2010–2016), Tom Rivett-Carnac (also previously United Nations) and co-hosted by Paul Dickinson (a sustainable economy expert). Christiana, Tom, and Paul call themselves 'stubborn optimists' throughout the podcast, now in its fifth season. In the weekly episodes, they discuss current issues, interview climate activists from different sectors and conclude with a musical interlude. The first episode was released in April 2019, with David Attenborough as the interviewee. In recent seasons, each episode concludes with a musical guest with an environmental song. *Sustainababble* advertises itself as a comedy podcast about the environment and sustainability, hosted by environmental charity workers 'Ol and Dave' (Oliver Hayes and David Powell, although neither use surnames in the podcast). The weekly podcast began in January 2015 and includes interviews with various experts. Snippets of dialogue from comedy series such as *Black Books*, *Monty Python*, and

The Simpsons are interspersed with its more earnest, discursive substance, lightening the tone.

We analysed the podcasts for rhetorical identification. Rhetoric is a discipline that evaluates the techniques and strategies of persuasive communication. As Arne Naess (1987: 35–41) suggests, environmentalism often dissuades people by couching its arguments in terms of sacrifice rather than personal fulfilment, rendering a rhetorical analysis of environmental communication invaluable for those seeking to understand its potential impacts. Kenneth Burke (1969) notably theorized that a process of 'identification' with a central tenet may arise if communicators establish commonalities with their audiences. We analysed the podcast episodes using Burke's rhetorical pentad: *the act* taking place, *the scene* or context of the act, *the agent* completing the act, *the agency* through which action takes place, and *the purpose* of the act (Borchers 2006: 153). The act is reinterpreted by the rhetor—the podcast host—according to the other elements of the pentad (Tonn et al. 1993). Gusfield (1989) suggests that a ratio imbalance between pentadic elements illuminates how they have been adapted to suit an act of communication. Our analysis demonstrates that podcasting (act) in the context (scene) of climate change introduces experts and celebrities (agents) who create narrative humour and emotional excitement (through their agency) to motivate listeners to make sustainable choices (purpose). We focused on all five points of the rhetorical pentad but discovered that agency through the format of the podcast was most compelling, offering both affordances and limitations.

Differences between emotion, mood, and disposition hold relevance to rhetorical agency—in our case, an ecocritical pathos identifiable across podcast episodes—and ergo to this chapter. When speaking of short-term emotional experiences, one ordinarily refers to 'reflex emotions': fleeting emotional responses instigated by one's situation. Over time, a sustained emotion can become constitutive of a 'mood'. 'Dispositions' are more ingrained emotional responses. They are effectively emotional shortcuts, habitual reactions to familiar tasks and circumstances, giving us emotional feedback regarding our predetermined goals, whether they are being met and whether we need to change our behaviour (Marcus et al. 2000).

The Eco Echo-Chamber

An important consideration for podcasters is their presumed audience and the rhetoric through which they are likely to be convinced. Communications researchers find that climate sceptics identify less with a perceived 'ingroup' than do believers (Bliuc et al. 2015: 227), and some research suggests that climate sceptics are also less certain of their own opinions (Poortinga et al. 2011). Moreover, those who do not believe climate change is a serious issue still support some emissions-reducing policy interventions, although (like their more committed peers) they are less supportive of 'environmental citizenship' that falls to the individual (Bernauer and McGrath 2016: 680). Thus, environmental communicators have differing kinds of work: to leverage climate uncertainty in persuading more people to the cause of demanding environmental policy changes and to leverage the group solidarity of believers to committed action.

We suggest that environmental podcasters aim their rhetoric at the latter, with the knowledge that the majority of their listeners will be believers rather than sceptics. Communicators must account for a context of 'information abundance' that is already available in climate debates: audiences actively seek out information that reinforces the prevailing ideology of their social networks (Bernauer and McGrath 2016: 682). It is therefore unlikely, in the case of environmental podcasts, that a simple matter of framing an argument will convince newcomers rather than galvanize an existing base of supporters (Fesenfeld et al. 2021).

Even within this group of supporters, diversity means that podcasters must cast a wide net of environmental communication. Narratives and emotions take a central place. According to Walter Fisher's narrative paradigm, humans are innate storytellers who can be convinced by narratives that demonstrate 'coherence and fidelity' (1987: 24). Fisher writes that narratives should 'ring true' by 'represent[ing] accurate assertions about social reality' (Fisher 1987: 105). Narrative consistency is important, so to be convincing, the values of the story should align with the listener's personal experience of causality in the world (Fisher 1987: 109). To appeal to their diverse audiences, podcasters vary their topics and interviewees weekly, increasing the chances for narrative fidelity to their listener's varied experiences.

A variety of interviewees who appear on the podcasts allow audiences to choose the ideas or people with whom they identify, and thus to act on ideas in the ways that best suit them as individuals. A repeated strategy is the respective introductions of celebrity and scientist guests, the disposition of the hosts toward those guests, and the positioning of the listener within a presumed, inclusive network of climate campaigning through entertainment media. *Sustainababble* and *Outrage + Optimism* use celebrity endorsement as a means of affective communication, often avowedly demonstrating that the environmental cause is substantial enough to cross a wide community of famous figures. Celebrity guests suggest a wider net of ingroup membership that includes people with power and means, appealing to group efficacy goals.

Hollywood guests such as *Outrage + Optimism*'s Jane Fonda, Ted Danson, Rainn Wilson, and Richard Curtis receive a more abbreviated introduction than other guests, and celebrities' contributions to the cause are often taken as self-evident. Contrariwise, when introducing environmental experts, the hosts of both shows set up the dependability of their expertise, explaining why the interviewee is a figure of note whose advice ought to be heeded. Celebrities offer a Postively valenced affect associated with the comforts of recognizability that is seldom available to environmentalists, even those renowned in their own communities. Expertise, meanwhile, can be correlated with their remit on agreeing to interview: to explain, in accessible terms, a difficult subject behind which has come years of research. Convincing an audience of the trustworthiness of a guest's conclusions may, in some cases, be easier than asking them to reason their way toward those conclusions via complex chains of expert analysis.

Another common identificatory strategy is to target ingroups, such as parents. For example, in *Sustainababble* (2021b), Ol and Dave interview Jen Gale about her book on eco-parenting. As she explains, 'I think we probably rapidly become more aware of our impact at exactly the same time as we've got no time, energy or head space to do anything about it'. Gale identifies with the broadly familiar experience of attempting to raise children in an environmentally sustainable way, using the collective pronoun 'we' to highlight collective failure. Her advice for how to help parents make sustainable changes is 'sharing your discoveries as you're doing them and your genuine excitement'. Similarly, in *Outrage + Optimism* (2021a), Tom interviews the children's author Oliver Jeffers, using a personal narrative to allow audiences to find points of identification with the podcast. He relays how they met online and began a pandemic-necessitated digital friendship that has given Tom 'A small but significant part of my optimism in the world' (*Outrage + Optimism* 2021a). They discuss how children can be part of conversations about the environment, the world, and our collective humanity. Jeffers suggests, 'A joke is always funnier when you tell it to someone else'. The advice of interviewees on both podcasts is followed by the podcasts themselves. Podcasters share their environmental discoveries with genuine excitement and joke with their listeners, emotionally engaging them to open conversations through areas of shared interest.

Our analysis identified the workload that the first-person collective pronoun 'we' takes on. The podcasts often address their audiences as collective groups; for example, in *Outrage + Optimism*, Christiana, Tom, and Paul interview Luisa Neubauer about her grassroots movement in Germany occupying offices and public spaces to picket every governmental party before an election to elicit promises of climate targets. She helps listeners overcome fears of insignificance in political action by seeking identification through the universal need for people to feel part of a community:

> There are millions of people organising themselves right now as we speak, or as someone listens to this. There are people out there who no matter how difficult the challenges are getting together and doing stuff. I consider myself to be so privileged and I see people fighting so hard every moment, and they rely on me and us to do so as well [...] There's this kind of ecosystem of hope where everyone knows I am this kind of hopeful person for someone else right now. (*Outrage + Optimism* 2021c)

Throughout the episode (*Outrage + Optimism* 2021c), the pronoun 'we' interchangeably refers to activists (*what 'we' can achieve*), adults (given the topic of youth protest and attitudes to climate, *'we' need to learn to listen*), those with hope (*'we' will prevail*), those without hope (*'we' sometimes feel climate anxiety*) and all human life (*'we' are facing an existential threat*). This casts a wider net of group cohesion, and also includes listeners in various emotional states, of varying emotional dispositions, and acknowledges mutability of moods. In effect, it acknowledges a range of feelings to bolster a sense of inclusivity.

Climate Change Comedy: Full of Oxymorons

One of the podcasts' strategies for listener retention is comedic identification. A central idea of Kenneth Burke is that guilt, either individual or collective, can lead to action for change. The hosts of both podcasts appeal to listeners' guilt about their roles in environmentalism, with the hope that this guilt could lead to a change in habits and further action. Comedy and optimism are important tonal approaches in both podcasts, giving audiences a sense of hopefulness necessary for change. Burke (1966) suggested that a group of people can feel 'categorical guilt', such as that felt by the podcast listeners who are a diverse group united in collectively seeking out an environmental podcast. This selective listening implies categorical guilt is already felt, and that listeners may already intend to learn how to become more environmentally sustainable and active.

In *Rhetoric of Motives* (1969), Burke explained that identification occurs in the space where two individuals have similar, or 'consubstantial', interests that can lead to an act of persuasion. Moving away from historical conceptions of rhetoric as a one-way persuasion by a communicator, Burke investigated ideas of identification and guilt that originated with the audience; that is, the ways that audiences *convinced themselves to be persuaded*. Burke wrote that messages can only be persuasive to audiences when they 'speak in the language of a voice within' (1969: 39). Environmental podcast listeners thus have agency in their own persuasion, listening to the podcasts because of an internal need to take more of a role in sustainability.

According to Burke, audiences could respond to categorical guilt through tragic redemption or comic enlightenment. While tragic redemption calls for a group to find a scapegoat for their guilt and cast them out (Burke 1969), comedy can be another tactic to assuage the group's guilt (Brummett 1984). As Carlson explains, 'The comic frame identifies social ills as arising from human error, not evil, and thus uses reason to correct them' (1986: 448). Borchers points out, 'One of the fundamental differences between tragic and comic redemption is the nature of the guilt [...] Rhetors pursuing the comic route focus on the *mistakes* of a fool in the rhetorical community' (2006: 158, original emphasis). By poking fun at these comic failures, audiences are encouraged to learn by safely and light-heartedly identifying themselves in the figure of the scapegoat. The comic scapegoat, according to Hugh Duncan, 'can be laughed at but he is also being laughed with. We are laughing at him to purge him—and ourselves—of folly' (1962: 395). The result is that audiences learn from the comic scapegoat's mistakes, providing them with enlightenment on their own actions and follies. Audiences then understand that a dominant schema for thinking about a situation may have been flawed, leading them to view the topic from another perspective.

The two podcasts analysed herein have clear tendencies toward the rhetorical approach of comic enlightenment. As *Sustainababble* claims on the 'About' page of its website, 'Your hosts Ol and Dave are supposed to be experts in how to save the planet. But why's everything got to be so confusing? Each week they try to get a little bit closer to enlightenment. They may fail' (*Sustainababble* 2022). Ol and Dave set themselves up as the comic scapegoats through which their audiences laugh at their environmental mistakes, draw associations with their own environmental responsibilities and learn how to correct their behaviours without feeling the associated burden of guilt.

Categorical guilt is the result of our fruitless pursuit of perfection (Burke 1966: 17–18). The idea of failed perfection is raised throughout the podcasts' interviews. For example, in a *Sustainababble* episode on green-washing, Ol and Dave interview Georgina Wilson-Powell on the use of palm oil as a stabilizer in peanut butter. Ol remarks, 'It's also a kind of reflection of how we've been marketed at, to think that anything that isn't perfect is a bit yucky' (Sustainababble 2021a). Wilson-Powell discusses her book on everyday environmental choices, explaining, 'people Google things and they're so overwhelmed with conflicting information'. Questions Ol and Dave raise include common queries such as which dairy-alternative milk is best for the planet (oat—soy and almond are worst) and what is the greenest pet (rabbits). Ol narrates his failed attempt to purchase a reusable coffee cup, describing its first outing:

> We went out for a lovely walk in Epping Forest on a beautiful, sunny, autumnal day and I lost my keep cup somewhere in the woods. So not only did I only get one use out of it, completely negating the environmental gains of having a reusable cup, I accidentally lobbed it in pristine nature. So, I'm really sorry Epping Forest and the planet in general. (*Sustainababble* 2021a)

As usual, the hosts set themselves up as the comic scapegoats and demonstrate that perfection is not a practical aim in environmental sustainability.

Modulating Eco-Anxieties Toward Collective Optimism

Both podcasts often return to the topic of eco-anxiety, through which expert interviewees also become identifiable and relatable. On *Sustainababble*, Clover Hogan defines eco-anxiety as 'Any feeling of fear or overwhelm or grief around the impending climate collapse or ecological breakdown' (*Sustainababble* 2020a). She cites research that this eco-anxiety is experienced by over 75% of young people in the UK. In these opening discussions of the episode, Ol and Dave address a source of categorical guilt for listeners: younger generations have been left with not only physical but also psychological distress because of the impact of older generations on the environment. They ask Hogan how she would respond to people who blame young activists such as Greta Thunberg for stirring up unnecessary anxiety. Hogan suggests, 'It speaks to a much deeper-seated fear that a lot of adults experience around the state of the world [...] No matter your age, you're going to be exposed to the reality of our situation' (*Sustainababble* 2020a). Together, they draw a connection between the different generations of listeners, turning the feeling of guilt into a collective area of identification. They then discuss ways of alleviating the collective guilt. Ol asks how he could talk to his children about climate change, admitting that he has not yet done so, and thus takes on the role of the comedic scapegoat. Hogan also assumes the mantle of the scapegoat, acknowledging that there is no perfect answer:

> We don't know if we're going to be able to solve the climate crisis. We want to be able to conclusively say that we're going to be able to rally the troops [...] but we fundamentally don't know. It's actually very healthy to develop a relationship with uncertainty from a young age. So, while you can't make promises about the future, you can make promises about how you're going to support your young person to do everything you can in your power to take action on the issues that matter and to support them to do the same thing. (*Sustainababble* 2020a)

Hogan's response offers a simple first step to listeners to help them alleviate their categorical guilt: have honest and open discussions with their families. In the final minutes of the interview, Hogan reiterates her own mistakes, despite her significant role as an environmental activist, thus heightening the identification with listeners:

> By the nature of living a privileged lifestyle, I am contributing to that problem. And that's a really difficult reality to wake up to. When I go to the supermarket, I am contributing to the climate crisis. When I go shopping on the high street, I am contributing to the climate crisis. And so of course this cultivates a sense of being powerless because you realize that you are complicit. (*Sustainababble* 2020a)

In addition to these points of identification, corralled toward a striving to improve our behaviour collectively, the hosts of each podcast take positions of deference to expertise. The *Sustainababble* hosts maintain a presence of affable, unthreatening, and jocular British masculinity through deferent phrases and vocal artefacts. When Dave and Ol introduce guests, they often signal their unworthiness in the face of either their guests' notoriety and eminent standing, or their accolades within a research or activist community. This reasoned deference signals an attitude presumed to be shared by hosts and audience, and that is a component of a tacitly promoted disposition toward authority figures who have earned authority by legitimate means. This disposition asks all non-experts to recognize their own limited knowledge in fields that involve years of study and research, yet must be balanced with opportunities for listener identification with specialists. In essence, this deference contributes to group efficacy goals: it demonstrates that the community has the support of reasoned experts, and in deferring to their expertise, we will together find solutions to our common problems.

In contrast, *Outrage + Optimism* host Christiana Figueres communicates more vocal excitability relative to her peers. Excitability, communicated through more frequent changes in vocal pitch and tone, includes the listener within an elevated emotional space created by the host, appealing to intimacy and animacy. Given its affective intensity, vocal excitability leverages the potential for emotional contagion to create a shared mood (c.f. Hatfield et al. 1993). Like humour and laughter, expressive vocalizations can be infectiously exciting. They have a call to inclusivity in their socially contagious nature, building on the capacity to inspire feelings of social closeness and bonding in podcast audiences. The prospect of activism becomes exciting rather than depressing and creates a tandem *ethos* of relationality, social inclusivity, and group cohesion.

A Balancing Act of Emotions

Environmental podcasts aim not to simply create an emotionally hopeful space during their runtime, nor retain that sense of hope in the individual after the podcast; the higher order goal is to shift a collective mood and contribute to a lasting dispositional change among a greater audience. Podcasters do not only target the listener's emotional state but what that listener will then bring to others. Activistic media producers must attempt to activate a continuum between short-term reflex emotions, moods in media and affective dispositions of a political nature. Producers have at their disposal emotion-inspiring resources such as music, vocal character, subject choice, and so on, which together can contribute to an overall mood. They then hope that immersion in this mood will lead to a disposition that is taken beyond the episode, which entails action in the world. A key question for podcasters is what mix of 'reflex emotions' is likelier to translate into longitudinal dispositions. We are interested in how those reflex emotions instigated by podcasters contribute to group solidarity, efficacy and the directedness of anger.

The title of *Outrage + Optimism* shrewdly identifies two emotions that aid in lifting us to action: anger at injustice and hope that activism will make a difference. Given the importance of solidarity as a predictor of collective action, another point of interest is how the relational feelings that bind a group, like sentimentality and shared levity, can contribute to group cohesion. Media producers can bind an audience in a collective identity by creating a sense of shared, appropriate emotional response to the subjects raised. As James M. Jasper (2011: 294) writes, 'Groups seem to be strengthened when they share reflex emotions in response to events and when they share affective loyalties to one another [...] with each one contributing to the other'. These contribute as affective motivators, yet it still matters what subjects these emotions are directed toward in building a collective identity and moving forward with the right action, rather than action that inflames community divisions without being targeted toward a policy goal. As such, Huber et al. (2020: 383) find that 'more empathetic communication of climate-related policy goals might help to increase acceptance and support for these policies'. A blend of appropriately directed anger, emotions that bind a group and empathy that widens the group is a difficult balancing act.

So, what does the evidence tell us about the effects of these differing emotions? Jasper (2011) notes that many decades of research have demonstrated the short-term emotional inducements of protest that can serve or constrain longer-term political goals. An individual's commitment to collective action can be predicted by several contributing factors, including 'social identification with (or commitment to) a relevant group, a sense of grievance or perceived injustice that is expressed as anger at opponents, and beliefs that the group can achieve its goals (group efficacy beliefs)' (Bliuc et al. 2015: 226–227). For political communicators, then, rhetoric that reinforces group identities and the efficacy of the group's action are paramount—in affective terms, anger and hope are both useful tools for inspiring activism. In the specific case of climate action, the balancing act of emotions has some of its own particularities, especially when stressing the existential threat of climate change, which can equally generate climate anxiety (Clayton 2020) and a paralyzing catastrophism, popularly known in the media as 'doomerism' (Truscello 2018). This is the belief that if the threat is as substantial and existential as reported, then it is insurmountable and climate action is futile. As Jasper (2011: 291) writes, 'In their pleas for support, activists must temper the pleasures of accomplishing an impact with a continued sense of fear, anger, and threat that demands continued action', and this is a difficult balance to achieve.

Building solidarity and selling hope (group efficacy) toward collective action are the rhetorical goals of environmental podcasts. Note that while anger directed toward an outgroup is a strong motivator for political action, intergroup conflict can also alienate those who are not already committed and further entrench the beliefs of both sides of a conflict (Bliuc et al. 2015). This puts at odds the two potential goals of environmental communicators, representing another affective balancing act: rage can inspire action among the committed but also frames that action in oppositional rather than collective terms, which can be a disincentive for those on the fence. It also matters who one perceives as their opponent. Where an outgroup must be identified to bolster climate action, it seems reasonable to suggest that it is more productive to present members of organizations *preventing* climate action as subjects worthy of ire rather than those sceptics who are convinced by lobby groups and political campaigns yet hold no position of power. At the same time, recent research finds that a more inclusive rhetoric that casts politicians and citizens as working together can help convince partisan audiences to support climate goals, likely because it increases notions of both group efficacy and inclusivity:

> emphasizing that the political elite is responsive to citizens' needs and preferences in a climate-specific context significantly increased climate concern, support for climate policy, and willingness to pay [...] However, in contrast to our expectations, cues that commonly appear in populist discourse that involve blaming politicians for their indifference towards citizens' preferences have no negative effect on policy support. (Huber et al. 2020: 382)

This again points to differing rhetoric required for differing audiences.

Both podcasts we analysed include guests who discuss the differing emotional languages of environmental communication and their effectiveness. In *Sustainababble*, Dave and Ol interview environmental communications expert Robin Webster about the most effective way to discuss climate change. She explains the importance of 'understanding people's value systems, listening to where they are [and] sometimes focusing on the connections between two people if you're having a conversation rather than focusing on winning some kind of fact-war about sea level rise' (*Sustainababble* 2020b). She also asserts that inaccessible scientific language is a deterrent for audiences, suggesting, 'It doesn't speak to normal human beings who are trying to talk about some kind of profound change to our society using all these really bizarre words that don't mean anything to anybody'. Ol and Dave are, similarly, inclined to play a klaxon sound if words like 'neoliberalism' are used on their podcast. Webster discusses the findings of her research, that communities need trusted messengers to deliver affective environmental communication: 'It's about providing the empowerment and support[ing] people within communities who can speak to people who are empowered to hear them as authorities, rather than trying to talk a little bit louder'. Indeed, the agency of audiences is important in urgent communications like those in environmental podcasts. Krista Ratcliffe argues rhetorical listening is 'a stance of openness that a person may choose to assume in relation to any person, text, or culture [...] the purpose being to negotiate troubled identifications in order to facilitate cross-cultural communication on any topic' (2005: 17). Again, by including a variety of interviewees from different global communities, the podcast hosts widen the scope of relationality and increase the chances that listeners will engage rhetorically or be empowered to listen.

While anger is often brought up as a motivator in the podcasts, the balancing power of hope is also frequently discussed. In *Outrage + Optimism*, Christiana, Tom and Paul interview activist and author, Rebecca Solnit, on the necessity of hope in environmental communication. Solnit argues:

> Americans around me have often suggested a kind of binary thinking: either we're confident everything will turn out fine in which case we don't need to do anything, or everything is going to hell and there's nothing we can do about it. And they're both justifications for passivity. And so, there's a tendency to think hope means you don't have to do anything [but] for me hope is an engagement with radical uncertainty. We don't know what's going to happen; we may be able to participate in shaping what can happen. (*Outrage + Optimism* 2020)

She expands on this idea by arguing that despair can be 'a luxury good and an excuse' and that hope and climate action must work together to be successful.

In sum, making audiences feel like part of a community and stressing the efficacy of collective action are key. Anger is trickier, as the entities to which anger is directed can be motivating for different audiences and can sometimes be at odds with inclusivity and group efficacy goals. Communicators walk a line between reasonable anger at obstructive entities, attempts to cast a wider net of group membership and to sell the group membership as effective (presenting an *ethos* that includes specific members with the power to create meaningful change, of relevance to podcasters' choice of guests). This is ultimately another reason for environmental communicators to know their audiences, and to know who their audiences perceive as ingroup members and adversaries, in order to choose the most inclusive and motivating rhetoric.

Mixed Emotions as Aural Motivation

Podcasters achieve multiple balancing acts by creating an emotive mood throughout an episode and, to be convincing, that mood needs to be characterized not by an overarching feeling—like rage—but mixed emotions that properly correspond to the varied experience of thinking about environmental issues, obstacles, and opportunities. For instance, engendering hopefulness is not as simple as sustaining a positively valenced mood throughout an episode: this would risk cloying disavowal not simply of any negative affect, but of hopelessness, exhaustion, nihilism, and all those aversive emotions and attitudes that audiences may sometimes feel but that are anathema to action. In a sense, podcasters must pay respect to feelings of hopelessness in order for hopefulness (and ergo group efficacy goals) to appear realistic. To respect feelings of hopelessness or climate anxiety occasions the prospect of adequately dealing with those emotions, treating them as subjects to explore rather than avoid.

'Mixed emotions', sometimes called 'complex emotions'—as distinguished from Paul Ekman's (1999) 'basic emotions'—are characterized by a blend of positive and negative valence (Larsen and McGraw 2011). For example, bittersweetness is a mixed emotion that combines happiness and melancholy, positive and negative feelings. Note that movement between the comforts of a shared mood, and the motivation of new threats, is also the stuff that drama is made of—movement between emotions is fundamental to storytelling. Stories can feel lifelike when those emotional experiences overlap. Storytellers and podcasters alike use resources including humour and the mixed emotions of music to 'lighten the mood' between segments of more dramatic information. To maintain gravitas, that mood and its intended translation to a disposition (in acts of political persuasion) must be balanced with suitably dark feelings re-establishing the stakes. This is because dispositions are more intuitively habitual, and so less likely to inspire action than a moral disposition that is then activated by attendance to new threats (Marcus et al. 2000). So, podcasters broadcast points that ought to distress their audience in order to activate surveillance of new courses of action, while at the same time comforting the listeners with a sense of group cohesion and efficacy.

We turn now to focus specifically on the ways in which music facilitates a blend of affect, and an emotional journey that serves its own persuasive goal: reminding one of climate threats and how they make us feel, while positioning those threats as affectively surmountable when shared. One trope of podcasts featuring narrative reportage is what we call the 'final act shift'. The final act shift, taking place after an episode's climax and leading into its dénouement, is marked by a retrospective change in the *pathos* of its scoring. Left-wing, humanistic, narrative-based podcasts have fairly consistent sonic and harmonic conventions: for instance, take the popular example of NPR's *This American Life*. The show will often use Rhodes or electric keyboards with tine emulators playing slow, rhythmic chord structures that resist resolution to the tonic to maintain a sense of narrative tension. To avoid overwhelming the spoken component, the music tends to be harmonic and rhythmic more than melodic—a melody is usually only suggested in the chord progression. The sound is often compressed, unembellished and dry (that is, with minimal reverb and delay). Together, this creates a sense of unresolved mystery, yet simultaneously retains the warmth and intimacy of a small room. The final act shift will introduce similar music, at this point including a resolution. The timbre will be similar; however more likely to be played on major chords with minimal extension, suggesting the comforts of resignation to a grander truth gleaned from the journey of the episode's narrative. While the story told may not proffer any easy answers to the socio-political problems of its object, this shift encourages listeners to accept the episode's insights, told via a satisfyingly dramatic storytelling arc, as some manner of useful progress.

Environmental podcasts, while they share many other conventions, do not quite have an identical luxury. Given their call to urgent action, arising from a severely existential dilemma, such an affective emphasis on resolution might seem to run contra to the activist agenda. *Sustainababble* and *Outrage + Optimism* largely skirt the dilemma by including less incidental music than other podcasts and punctuating their more downbeat dialogue with an upbeat, percussive theme tune. By the same token, some manner of sentimentality balanced with outrage—rather than sentimentality or ire in isolation—appears to be understood by ecocritical podcasters as appropriately motivating. The use of music is directly addressed within *Outrage + Optimism*'s 'Stubborn Optimist's Playlist' episodes (*Outrage + Optimism* 2021b), in which they pay homage to musical artists featured as interviewees in earlier instalments. Again, they choose upbeat, and occasionally sentimental music, with affect less often characterized by the resigned and intimate cues described above.

Music and collective movements, such as marching through the streets, have long been noted for their ability to mobilize a sense of communal joy and investment in group identity (McNeill 1995). As physical movement is unavailable to the podcaster, music effectively has big shoes to fill. Music choice in environmental podcasts provides a good case in the balance of mixed emotions required for those podcasts to be effective: often, music provides unambiguously positive affect to offset against downbeat subject matter within the dialogue. Yet, at times music is also used to suggest a range of other emotions that come together to create a more affectively mixed experience; one that admits sadness or anger, and when balanced with discussion of climate issues, feels more akin to the ups and downs of climate activism. As Christiana Figueres puts it when introducing 'The Stubborn Optimist's Playlist Vol. 1' (*Outrage + Optimism* 2021b), 'we have a lot of fun, and sometimes we go into deep pain—so we go through the whole gamut of feelings, from outrage all the way to optimism and back'. The one thing that is conspicuously absent from their scoring toolbelt, however, is musical resolution.

This lack of resolution harkens back to the form of the podcast and foundational ideas of affective rhetoric. Since Sarah Koenig narrated the first season of the ground-breaking true crime podcast, *Serial*, in 2014, the format has often offered listeners a sustained mystery- or puzzle-based narrative. Rhetorically, Edwin Black argues that concluding an argument forecloses it as a source of communication and dialogue (1970: 109). Not so for environmental podcasts, which contain the ongoing narrative context of climate change and the evolving global political and social responses. A foreclosed argument without complicating detail would not invite listeners to move to action and activism.

Conclusion

Podcasts such as *Outrage + Optimism* and *Sustainababble* present listeners with a complex and varied approach to environmental sustainability. The podcasters adapt their urgent environmental communication to the format afforded by the podcast medium in compelling and clever ways. Our analysis of the two podcasts highlighted the difficult balance of emotions—particularly anger and hope—through personal narratives and music.

Addressing their rhetoric to an audience of listeners aware of their role in the climate crisis, but from otherwise varied social, cultural, and global communities, the hosts present an abundant offering of interviews with experts and celebrities of different backgrounds. These interviewees may appeal to diverse audiences through niche identifications, such as parents, fans, or generational groups. Podcast hosts and interviewees consistently set themselves up as comic scapegoats who acknowledge and laugh at their environmental follies, demonstrating how audiences can rally from similar mistakes. The conclusions of episodes offer no ultimate resolution but invite audiences to make their own resolutions through individualized environmental action.

References

Bernauer, Thomas and McGrath, Liam F. (2016), 'Simple Reframing Unlikely to Boost Public Support for Climate Policy', *Nature Climate Change,* 6:7, pp. 680–683.

Black, Edwin (1970), 'The Second Persona', *Quarterly Journal of Speech,* 56, p. 112.

Bliuc, Ana-Maria, McGarty, Craig, Thomas, Emma F., Lala, Girish, Berndsen, Mariette and Misajon, RoseAnne (2015), 'Public Division about Climate Change Rooted in Conflicting Socio-Political Identities', *Nature Climate Change,* 5:3, pp. 226–229.

Borchers, Timothy (2006), *Rhetorical Theory: An Introduction,* Belmont, CA: Thomson Wadsworth.

Brummett, Barry (1984), 'Burkean Comedy and Tragedy, Illustrated in Reactions to the Arrest of John DeLorean', *Central States Speech Journal,* 35:4, pp. 217–227.

Burke, Kenneth (1966), *Language as Symbolic Action: Essays on Life, Literature and Method.* Berkeley: University of California Press.

Burke, Kenneth (1969), *A Rhetoric of Motives*, Berkeley, Los Angeles, and London: University of California Press.

Carlson, A. Cheree (1986), 'Gandhi and the Comic Frame: "Ad Bellum Purificandum"', *Quarterly Journal of Speech,* 72:4, pp. 446–455.

Clayton, Susan (2020), 'Climate Anxiety: Psychological Responses to Climate Change', *Journal of Anxiety Disorders,* 74, n.pag.

Duncan, Hugh Dalziel (1962), *Communication and the Social Order*, New York: Bedminster Press.

Ekman, Paul (1999), 'Basic Emotions', in T. Dalgleish and M. Power (eds), *Handbook of Cognition and Emotion,* Sussex: John Wiley & Sons, pp. 45–60.

Fesenfeld, Lukas P., Sun, Yixian, Wicki, Michael and Bernauer, Thomas (2021), 'The Role and Limits of Strategic Framing for Promoting Sustainable Consumption and Policy', *Global Environmental Change,* 68, pp.1–12.

Fisher, Walter (1987), *Human Communication as Narration: Toward a Philosophy of Reason, Value and Action*. Columbia: University of South Carolina Press.

Gusfield, Joseph R. (1989), 'Introduction', in K. Burke, *On Symbols and Society*, Chicago and London: The University of Chicago Press, pp. 1–49.

Hatfield, Elaine, Cacioppo, John T. and Rapson, Richard L (1993), 'Emotional Contagion', *Current Directions in Psychological Science*, 2:3, pp. 96–100.

Huber, Robert A., Fesenfeld, Lukas and Bernauer, Thomas (2020), 'Political Populism, Responsiveness, and Public Support for Climate Mitigation', *Climate Policy*, 20:3, pp. 373–386.

Jasper, James M. (2011), 'Emotions and Social Movements: Twenty Years of Theory and Research', *Annual Review of Sociology*, 37, pp. 285–303.

Larsen, Jeff T. and McGraw, Peter (2011), 'Further Evidence for Mixed Emotions', *Journal of Personality and Social Psychology*, 100:6, pp. 1095–1110.

Marcus, George E., Neuman, W. Russell and MacKuen, Michael (2000), *Affective Intelligence and Political Judgment*, Chicago: University of Chicago Press.

Markman, Kris M. and Sawyer, Caroline E. (2014), 'Why Pod? Further Explorations of the Motivations for Independent Podcasting', *Journal of Radio & Audio Media*, 21:1, pp. 20–35.

McClung, Steven and Johnson, Kristine (2010), 'Examining the Motives of Podcast Users', *Journal of Radio & Audio Media*, 17:1, pp. 82–95.

McNeill, William H. (1995) *Keeping Together in Time: Dance and Drill in Human History*, Cambridge: Harvard University Press.

Naess, Arne (1987), 'Self-Realization: An Ecological Approach to Being in the World', *The Trumpeter*, 4:3, pp. 35–41.

Outrage + Optimism (2020), '66: Embracing Uncertainty with Rebecca Solnit', *Global Optimism*, 21 August, https://www.outrageandoptimism.org/episodes/embracing-uncertainty-rebecca-solnit?hsLang=en. Accessed 1 March 2022.

Outrage + Optimism (2021a), '88: Living on Planet Earth with Oliver Jeffers', *Global Optimism*, 11 February, https://www.outrageandoptimism.org/episodes/living-planet-earth-oliver-jeffers?hsLang=en. Accessed 1 March 2022.

Outrage + Optimism (2021b), '95: Stubborn Optimist's Playlist Vol. 1', *Global Optimism*, 1 April, https://www.outrageandoptimism.org/episodes/stubborn-optimists-playlist-vol-1?hsLang=en. Accessed 1 March 2022.

Outrage + Optimism (2021c), '119: Dealing with Climate Grief with Luisa Neubauer', *Global Optimism*, 7 October, https://www.outrageandoptimism.org/episodes/dealing-with-climate-grief-luisa-neubauer?hsLang=en. Accessed 1 March 2022.

Poortinga, Wouter, Spence, Alexa, Whitmarsh, Lorraine, Capstick, Stuart and Pidgeon, Nick F. (2011), 'Uncertain Climate: An Investigation into Public Scepticism about Anthropogenic Climate Change', *Global Environmental Change,* 21:3, pp. 1015–1024.

Ratcliffe, Krista (2005), *Rhetorical Listening: Identification, Gender, Whiteness*. Carbondale: Southern Illinois University Press.

Riordan, Kellie (2018), 'Commentary: The Mobile Phone as the New Transistor Radio', in S. Park, C. Fisher, G. Fuller and J. Y. Lee (eds), *Digital News Report: Australia,* News and Media Research Centre, University of Canberra. https://doi.org/10.4225/50/5b21a3c6f196a.

Sustainababble (2020a), '185: Eco-Anxiety', UK, 10 August, http://www.sustainababble.fish/?p=1123. Accessed 1 March 2022.

Sustainababble (2020b), '186: Talking Climate', UK, 17 August, http://www.sustainababble.fish/?p=1128. Accessed 1 March 2022.

Sustainababble (2021a), '199: Is it Really Green?', UK, 10 January, http://www.sustainababble.fish/?p=1207. Accessed 1 March 2022.

Sustainababble (2021b), '207: Eco-Parenting', UK, 14 March, http://www.sustainababble.fish/?p=1262. Accessed 1 March 2022.

Sustainababble (2022), 'About', http://www.sustainababble.fish/?page_id=2. Accessed 1 March 2022.

Tonn, Mari Boor, Endress, Valerie A., and Diamond, John N. (1993), 'Hunting and Heritage on Trial: A Dramatistic Debate Over Tragedy, Tradition, and Territory,' *Quarterly Journal of Speech,* 79.2, pp. 165–181.

Truscello, Michael (2018), 'Catastrophism and Its Critics: On the New Genre of Environmentalist Documentary Film', in J. Jagodzinski (ed.), *Interrogating the Anthropocene*, Palgrave Macmillan, pp. 257–275.

PART II

Ecoaesthetic Practice & Analysis

CHAPTER 6

A Close Reading of Climate-Related Art: Aesthetics and Creative Engagement with the Structural Causes of the Climate Crisis

Ulrike Hahn

Introduction

Climate-related art warrants a deeper look because it has the potential to create meaningful engagement with the complex topic of climate change. This chapter focuses on a select number of artworks to receive in-depth, rich, and interpretive insights. A close reading was done for the following four artworks: Ursula Biemann's *Deep Weather* (video essay, shown in gallery/museum context and online), Diane Burko's work *Summer Heat 1 & 2* of her *Seeing Climate Change* exhibition (painting, shown in gallery/museum context and online), Olafur Eliasson

U. Hahn (✉)
Arts and Culture Studies Department, Erasmus University Rotterdam, Rotterdam, The Netherlands
e-mail: hahn.artvisory@outlook.com

D. Binns and R. Najdowski (eds.), *Confronting the Climate Crisis*, Palgrave Studies in Media and Environmental Communication,
https://doi.org/10.1007/978-3-031-89606-4_6

and Minik Rosing's *Ice Watch* (installation, public space, photography and video material shown online and in gallery/museum context), and Jason deCaires Taylor's underwater sculpture *Vicissitudes* (installation, public space, photography shown online and in gallery/museum context).

These particular artists and artworks, shown in both public spaces and galleries/museums, were chosen for their diversity of mediums ranging from video art, painting to sculpture and installation. Two works (*Deep Weather, Summer Heat 1 & 2*) are representational (i.e. focusing on visual representation), and two works (*Ice Watch, Vicissitudes*) are experiential (i.e. one experiences the work with all their senses).

The close reading focuses on the aesthetics, if and how the works engage with structural causes and/or consequences of the climate crisis, and if the artwork contains an activist message.[1] The following research question was posed: *How do visual climate-related artworks employ aesthetics and engage with politics and structural causes of the climate crisis?*

Instead of a formal analysis (which investigates art on its own while social factors are mostly left aside), this artwork analysis looks at both the art and social factors and thus follows an interdisciplinary approach between sociology of the arts and the humanities. The present artwork analysis does not aim for generalization but for particularizing (Alexander 2021). It is an illustrative account of climate-related art, not an exhaustive mapping of all contemporary art dealing with the climate crisis.

Analysis

Ursula Biemann: Deep Weather (2013)

Description of the Artwork

The work *Deep Weather* (2013) by Swiss artist Ursula Biemann (born 1955) is a ca. 9-minute-long video essay. It has been shown in various places, for example at the Prada Foundation in Venice (2023), in the Center for Contemporary Art and Ecology, RADIUS, in Delft (2022), and at the Montreal Biennale (2014). At the time of writing this chapter, the work can be accessed on the website of the artist (Geobodies n.d.). The video essay is narrated by a whispering voice as white text appears

[1] Many more aspects would be relevant to consider but the analysis needs to be limited to a select few to keep it feasible.

on the screen, overlaying still images or video sequences. The video essay is divided into two parts, a ca. two-minute-long part entitled 'Carbon geologies' and a ca. 7-minute-long part entitled 'Hydrogeographies'.

After a short sequence showing cargo vessels moving along a shoreline, the camera pans across a river running through tar sand fields, seen from a bird's eye view. The irregular shapes of the fields and the water (often reflecting the cloudy sky) are contrasted by the straight lines of transport roads and construction works that cut through the landscape. Machines can be seen, which have carved their way into the landscape, and an industrial landscape out of which smoke is emitted (Fig. 6.1).

The whispering voice makes clear that the images have been shot in Alberta, Canada, more specifically around the Athabasca River area. The voice, which can be heard together with a sound of whooshing wind, explains that oil miners take out tar sands in an area as large as England. To transit to the second (and much longer) part of the video essay, the narrator connects the mining of tar sands to the rising sea level: 'For a hundred more years there is enough stuff for heavy fuel that will bring toxic clouds over the boreal woods and continue to warm and swell the Seas. No longer to be witnessed but elsewhere, in equatorial zones'.

At 2:20 min, the video changes to the second part, 'Hydrogeographies' (Fig. 6.2). The location switches from Canada to Bangladesh, and the camera changes from aerial view to eye level. Instead of the sound of wind, the sounds of people who are transporting and placing sandbags at the shore to make land can be heard. These full-screen shots are alternated with split-screen sequences showing a person on one half of the screen while the other half shows landscape scenes (water and a strip of land, no people). In each of the split-screen scenes, different people stand, directly

Fig. 6.1 Ursula Biemann, *Deep Weather*, 2013. Video essay. © Ursula Biemann. Video still of the aerial view provided by the artist

Fig. 6.2 Ursula Biemann, *Deep Weather*, 2013. Video essay. © Ursula Biemann. Video still of the split-screen sequence provided by the artist

looking at the camera, and thus the viewer (Fig. 6.2). Only in these split-screen scenes, the whispering narrator (with the text superimposed on the image and a similar sound of whooshing wind to the first part of the video) explains the impacts on the area and the activities taken by the people.

Existing Coverage and Artwork Analysis

Structural Causes Around the Athabasca River Area

While the artist is based in Switzerland, Biemann is involved in field-work in remote areas in many of her works, co-creating projects with local communities and connecting scientific and indigenous perspectives. For *Deep Weather*, she went on field visits to Northern Canada and the Ganges Delta to document the territory (Ouellet 2020). This is why I position Biemann's work as part of Galafassi et al.'s (2018) second movement of climate-related art, which focuses on co-creation.

The choice of the word weather in the title of the video essay may seem peculiarly chosen at first, as climate may appear a more appropriate term when talking about climate change. However, Biemann's combination of weather with deep makes the connection to deep time. Biemann (2016: 10) notes in a text about the work: 'As we scoop out fossil and mineral matters from deep geological strata into the daylight and out into the atmosphere, climate change makes us think in deep time'. Deep time can be understood as 'distant times beyond the scope of human experience' (Ginn et al. 2018: 214). Therefore, Biemann combines a word that is about the short-term (weather) with a word that has a long-term, distant time connotation (deep). This unification appears to be a contradiction, but it could also hint at the existence of consequences taking place already in the present (e.g. in Bangladesh), while more impacts will be experienced in deep time.

Biemann is well aware of the challenges of representing climate change, referring to philosopher Timothy Morton who sees it as a hyperobject, something that is enormously spread in space and time in relation to humans (Morton 2013). But she also finds that while we are not able to portray climate change, 'we can read it videographically through such sceneries' (Biemann 2016: 11). Her work shows both causes in the Global North (Canada) and consequences in the Global South (Bangladesh) (but also in Canada). It demonstrates how the impacts of a local ecology impact a remote location. Thus, it has planetary dimensions.

The video essay attributes responsibility by visually showing that 'Carbon geologies' (the extraction of tar sand) cause 'Hydrogeographies' (rising sea levels, cyclones). The whispering voice, which provides supporting information about the causes and effects, makes this unmissable. The sound of whooshing wind further supports this atmospheric connection (the atmosphere that brings greenhouse gases across the globe) between the two remote sites. Thus, *Deep Weather* addresses systems ecology. Not in the sense of many works described by Demos (2016) which *implement* systems ecology as part of the work, rather Biemann's work *represents* interconnected systems such as forests, oil and water ecologies, environmental dynamics and social action, local and planetary systems, and the Earth system at large.

Politics of Environmental Justice

Deep Weather is very clearly about climate change. Whereas there is no explicit activist call voiced in the video essay, the artist's biography makes clear that she does have a connection to activism. Biemann co-initiated a collective, World of Matter, which 'considers visual source material a valuable instrument for education, activist work, research, and raising general public awareness' (World of Matter n.d.: para. 3). Even though a concrete solution is not offered in the video, it seems evident that tar sand extraction cannot continue in the way it has been operating.

Moreover, the work can be seen as developing a political ecology. Biemann's use of text makes the political ecology even more clear and specifically a politics of environmental justice. This is because it examines causes and effects across the globe with attention paid to power relations and people, which has been described as particularly important for art that addresses climate change (e.g. see Demos et al. 2021; Yusoff and Gabrys

2011). For example, the video essay suggests that there is an imbalance between the sites of tar sand extraction (Canada) and sites of exhaustion (Bangladesh). While the industry in Alberta profits from the tar sand extraction, the Bangladeshi people suffer from rising sea levels. Moreover, the short duration of 'Carbon geologies' compared to the second part on 'Hydrogeographies' may imply that the safe distance from the mining of tar sands is only of short duration and that the impacts will be felt much longer, particularly by vulnerable communities. Climate change further exacerbates inequalities. However, the Bangladeshi people should not be seen as merely disempowered as they join forces trying to adapt to the dire situation.

From the Industrial Sublime to Eco-Realism

The 'Carbon geologies' segment portrays the causes of climate change from a bird's eye view, excluding human presence and creating a sense of detachment. Aesthetically, the images of the 'Carbon geologies' part remind of an industrial sublime, in which enormous human-made infrastructures are aestheticized. In Biemann's video, this is only one part of the work. In contrast to the first part, 'Hydrogeographies' appears closer and more confronting because it involves people (Bangladeshi citizens) directly facing the camera (during the split-screen scenes) and shown working at the shore.

Moreover, in the second part of the video, landscape scenes in split-screen appear in tones of grey and blue, visualizing the power of the sea, of nature, in which humanity does not have a place as no people are visible. This mirrors the river scenes in the first part, in which neither workers nor First Nations are portrayed. Biemann (2016: 11) talks about a 'post-humanist realism', and Ouellet (2020: 128) calls Biemann's approach 'eco-realist'. These are not future-oriented or dystopian scenes, but contemporary, realistic conditions.

The aesthetic choices and Biemann's portrayal of structural causes and her politics of environmental justice can be related to Kagan's (2011) concept of an aesthetics of sustainability. The work shows a sensibility to the complexity of climate change causes and effects, and it moves beyond a simplified, individualized, or localized perception. Moreover, it demonstrates a sensibility to connected patterns by 'unravel[ling] the Earth's ecologies as an interconnected system where its crises cannot be looked at in a vacuum', as the exhibition brochure of RADIUS stresses (RADIUS 2022: 10).

Diane Burko: Summer Heat 1 & 2 (2020)

Description of the Artwork
The colours are vivid in the exhibition room of *Seeing Climate Change* at The American University Museum in Washington DC. The exhibition features a visceral, seductive beauty that starkly contrasts with the devastating topic that is its focus. Many of the works are abstract, but based on data or maps. For example, the work *Summer Heat 1 & 2* (2020) is a large (84 by 162 inches; 213 by 411 cm) landscape format work, mixed media on canvas, which consists of several panels (Fig. 6.3).

At the left side, two smaller, approximately square panels are positioned on top of each other. The top left panel shows a world map, with the continents coloured black and the water in shades of blue. While the shape and size of most continents seem to follow approximately common cartographies, Antarctica on the bottom is featured as a large black mass, in front of which a chart is positioned. The chart depicts carbon dioxide concentration in ppm (parts per million) over time, however, the chart is difficult to read as it nearly vanishes into the black Antarctic territory. On top of the chart, the number 415,26 ppm can be read.

The lower left panel is mostly painted in red and black colours and appears to show a fire seen from eye level that is covering the sky, of which only a small part of blue is left in the top of the panel. A map of

Fig. 6.3 Diane Burko, *Summer Heat 1 & 2*, 2020. Mixed media on canvas. 213 cm × 411 cm. © Diane Burko Studio

Australia (in red) in the right bottom corner of this panel and the shape of a kangaroo next to a burnt tree that can be discerned within the fire suggests that it may be a bushfire in Australia. In the bottom part of the panel, on a black surface (possibly burnt earth), another chart can be seen which seems to be showing carbon dioxide concentrations again. The number 415,26 ppm is visible as in the top left panel.

On the right of the two square panels, a much larger panel again combines a map with a chart and parts that seem to evoke some eye-level view, this time of a water surface with ice. The map (Italy and Sicily are clearly identifiable) is coloured in shades of red. Underneath the map are shades of blue and white. On the right is another chart, this one seems to represent increasing temperatures over time. The blue (water-coloured) paint of the water surface with ice is dropping down in the bottom right section, contrasting the rising graph.

Existing Coverage and Artwork Analysis

Bearing Witness to the Impacts of Climate Change

The title of the exhibition, *Seeing Climate Change*, could refer to science's attempts to see and visualize climate change, or to the artist bearing witness to the impacts of climate change, such as rising temperatures, melting ice, endangered reefs, and wildfires, which she makes visible in her paintings.

Like Biemann, Burko has travelled to many sites that she documents (Burko et al. 2021). Both artists combine aerial and eye-level views in their works and show effects and causes that are simultaneously happening. In Biemann's video work, a causation is established, yet both phenomena (sea level rising in Bangladesh, tar sand extraction in Canada) also occur simultaneously. In Burko's painting, the causes and impacts are meshed together, and everything appears to happen at the same time. For example, in the smaller panels, the fires, which are triggered by increasing temperatures due to a rising carbon dioxide concentration, in turn release carbon dioxide, further exacerbating the situation. In the larger panel, temperatures are rising, glaciers are melting, and water is swelling.

While Biemann focuses on the extraction of tar sands in Northern Canada and its effect on the communities in Bangladesh, Burko's work is more broad, concentrating on carbon dioxide emissions and the global impact. Biemann shows specific local environmental and social systems (with planetary dimensions), and Burko represents natural systems more generally and shows scientific figures.

Art and Activism

There is no explicit written political, activist message within the work *Summer Heat*. Yet, similar to Biemann, Burko's work still leaves no doubt that it is about climate change. Through its colours, maps, graphs, and title, it becomes clear to the viewer that the work addresses melting glaciers, rising temperatures, CO_2 concentration, and fires. Through its visual language, the painting also subtly communicates what is necessary: lower carbon dioxide emissions.

Like Biemann, Burko engages in activism around her work. Burko considers herself an activist, and she is heavily involved in public engagement through lectures and interviews (Burko 2021). She aims to create environmental awareness and 'activist engagement among her audiences' (Burko et al. 2021: 20). Therefore, the activism happens in the form of events around the work but is not explicitly voiced in the work itself. Burko is moreover a feminist activist. For example, she co-founded the Women's Caucus for Art (WCA) and contributed to the establishment of the Distinguished Feminist Award.

Seductive Beauty and Terrifying Sublime

There are analogies between Burko's (born 1945) abstract work and the tradition of the abstract sublime, which was described by art historian Robert Rosenblum as the sublime in complete abstraction (Rosenblum 1961). Yet, there are also differences to the abstract sublime because Burko combines her art with maps, graphs, and symbols. The collaboration between art and science is highly important to Burko (Burko 2021). Her work is thus not completely abstract, but rather a combination of art, science, and abstraction—a referential abstraction instead of a complete abstraction (Burko et al. 2021).

The sublime plays a role in both Biemann's and Burko's work, yet in different ways (the industrial sublime in Biemann's *Deep Weather*, and the seductive beauty and terrifying sublime in Burko's *Summer Heat*). Moreover, both artists do not only use the sublime, but they also employ other aesthetic strategies (i.e. Biemann's eco-realism, Burko's referential abstractions and use of maps and graphs). In the exhibition catalogue (Burko et al. 2021), Norma Broude, who is curator along Mary D. Garrard, addresses the tightrope that sublime aesthetics walk. It has the potential to draw the viewers in, compel them, but it can also numb them. Burko is described as successfully navigating this challenge.

Burko address various other topics beyond melting ice in their art. Moreover, one of the roles of the arts is to document or visualize the impacts of climate change, some of which are happening right now on the planet. What remains important is paying attention to power relations, people and politics, inhabitants, and their voices (Yusoff and Gabrys 2011).

The next point of analysis now turns to the environmental impact *of the work itself*. In particular, the transport of the ice blocks from Greenland to European cities has been criticized in the press as observed by the organization Julie's Bicycle (2019). The tenor of some of these press reports is: *Practice what you preach*. Together with the NGO Julie's Bicycle, Studio Olafur Eliasson has investigated 'the impact art making can have on the environment' (Bottrill et al. 2021: 23). According to their assessment, *Ice Watch London* has resulted in 55 tonnes of CO_2e,[2] most of which is caused by the freight (movement of the artwork). This equals '52 people flying return from London to witness the icebergs melting' (Bottrill et al. 2021: 23). This insinuates that it is, environmentally speaking, much better to have the icebergs come to the people in Europe, instead of having thousands of people fly out to see them melt on-location. Eliasson's work was strategically shown during IPCC and COP meetings. Thus, one can compare the effect of the work on people and its environmental costs (CO_2), and some may argue that the effect is worth the costs.

The Ice Parliament and COP Meetings

The installation itself does not contain any explicit activist or political message. In fact, one can enjoy it without knowing its intended meaning. But implicitly the message is 'that the source of the ice, the parent glaciers in Greenland, were also declining due to climate change' (Jackson 2015: 484). Rosing explains that the amount of ice that was installed in the Copenhagen edition (100 tonnes) 'is equal to what melts every 1/100 of a second' in the Greenland icesheet (Underground Channel 2014). In addition, the timing and locations of the installation make its activist claim clear: the first edition in Copenhagen coincided with the publication of an IPCC report, the second edition was installed in Paris in 2015 where COP21 took place (Hornby 2017), and the third edition was installed in 2018 in London during the same time as COP24 was held in

[2] CO_2e refers to carbon dioxide equivalent.

Katowice (Julie's Bicycle 2019; Ice Watch London n.d.). In Copenhagen, the installation was located in front of City Hall, in Paris in front of the Pantheon (which is the resting place of many famous people who had a large impact on French society) and in London in front of Bloomberg's European headquarters (Bloomberg Philanthropies supported the work).

Eliasson has also made explicit, in interviews and on the websites about *Ice Watch*, that he hopes to create engagement in the form of climate action (Ice Watch London n.d.). For him, the ice bears resemblance to an ice parliament, in which agreements need to be made (Olafur Eliasson n.d. a). The artist's call for climate action, moreover, underlines that he is interested in changing the status quo. While *Ice Watch* does not offer solutions as part of its immediate experience, it does provide resources separate from the artwork (Ice Watch London n.d.). This activism that takes place aside from the work is comparable to Burko's approach.

The Artwork Is the Space That the Circle of Ice Invents
Olafur Eliasson (born 1967) and Minik Rosing (1957) both describe the melting of the ice sheets of Greenland as an abstract phenomenon because it occurs in remote locations and is rarely directly observed (Olafur Eliasson n.d. a). Moreover, as Eliasson points out, the climate debate (e.g. at the COP meetings) and scientific data are also abstract (UN Climate Change 2015). *Ice Watch* is said to turn this abstract phenomenon into a 'personal and up-close experience' (Roosen et al. 2017: 17). *Ice Watch* is an immersive and participatory work (Hornby 2017) because it is 'a direct and tangible experience of the reality of melting arctic ice' (Olafur Eliasson n.d. a: Information section, para. 1), and because it allows spectators to take part in it. Eliasson stated that he does not consider the circle of ice blocks as the artwork. Instead 'it is the space it invents' (Zarin 2015: para. 2). Thus, the work depends not only on the weather conditions but also on the space and the person(s) participating in it.

The ice blocks seem alienated when lying in the streets of these three major European cities and are usually not experienced at that scale by populations of European countries. Yet, by melting, they become part of the cities. People can experience the ice installation on several levels, with or without necessarily knowing the implicit message. There may be an urgency felt because of the elements of the artwork (the ice, the clock like a countdown or ultimatum): we are '*at the eleventh hour*' regarding climate change or action. But people may also engage with the blocks without thinking about climate change. In time lapses of the ice installation, a range of people can be seen engaging with the work. Expressions of laughter, excitement, awe, contemplation, and reflection can be observed in peoples' faces.[3] Some touch the ice, some hug it, some listen to its sounds, some engage with it alone, others in interaction with fellow spectators. Perhaps there is an empathy for the melting ice blocks but also an empathy for the aesthetics, as they are disappearing. Through this documentation of the work online, there is also a representational part to the experiential installation and even people who have not experienced the work on-site may be able to imagine how it must be to experience the ice blocks.

A study by Sommer and Klöckner (2019) investigated the perception of 37 artworks, including *Ice Watch*, at ArtCOP21 in Paris in 2015 (total number of questionnaire respondents: 874, number of respondents for *Ice Watch*: 33). Based on the responses and a cluster analysis, *Ice Watch* was assigned to the cluster labelled by the researchers as the challenging dystopia: art that portrays dystopian scenarios in a dark sublime. Further perception studies with audiences who have experienced *Ice Watch* are needed to assess what they think about the work, if they feel the urgency for climate action, or if they perceive the work as foreclosing the future.

Jason deCaires Taylor: Vicissitudes (2006)

Description of the Artwork

Vicissitudes is a permanent site-specific underwater sculpture located at 5 meters depth off the coast of Grenada by Jason deCaires Taylor (born 1974), a British sculptor with Guyanese roots. Twenty-six life-sized figures of people stand in a circle, outward facing, and holding hands (Fig. 6.5).

The sculpture is made of pH neutral cement material, which does not contain harmful chemicals, is low carbon and long-lasting (Underwater Sculpture n.d. a). This material turns the sculpture into an artificial reef, encouraging the growth of marine life to continuously transform the artwork (Fig. 6.6).

In earlier images of the work (see Fig. 6.7), the sculpted figures of children are more easily discernible. Their facial features, closed eyes, modern clothing, and bare feet are clearly visible. Initially, the sculpture was a grey

[3] See the photos and videos on the artist's website (Olafur Eliasson n.d. a).

Fig. 6.5 Jason deCaires Taylor, *Vicissitudes,* 2006. Underwater cement sculpture installation depicting a circle of human figures. Photo by Jason deCaires Taylor. © Jason deCaires Taylor

Fig. 6.6 Jason deCaires Taylor, *Vicissitudes,* 2006. Underwater cement sculpture covered in marine life. Photos by Jason deCaires Taylor. © Jason deCaires Taylor

colour, unlike the later brown, purple, and pink colouring due to marine growth (Fig. 6.6). Additionally, the figures are duplicates; there are 14 and 12 identical children.

Vicissitudes is part of 75 works of Taylor's Molinere Underwater Sculpture Park[4] in the Molinere Beauséjour Marine Protected Area, off the west coast of Grenada, an island in the Caribbean Sea. This underwater sculpture park was installed in 2006 and is seen as the first of its kind (Underwater Sculpture n.d. b). In the Molinere Bay, Hurricane Ivan led to significant damage in 2004 and further intensified problems of this ecosystem. Therefore, according to Taylor, the underwater sculptures offered a new foundation for the growth of marine life (Underwater

[4] One of Taylor's many underwater sculpture parks and museums.

Fig. 6.7 Jason deCaires Taylor, *Vicissitudes,* 2006. Underwater cement sculpture installation depicting a circle of human figures standing on a circular platform. Photo by Jason deCaires Taylor. © Jason deCaires Taylor

Sculpture n.d. b). The sculptures are located at depths ranging from five to eight meters, and can be reached by 'scuba diving, snorkelling and glass bottom boats' (Underwater Sculpture n.d. b: para. 3). The park is promoted through the artist website, social media (with a substantial followership), tourism sites, and experience reports.

Existing Coverage and Artwork Analysis

Impacts and Causes

Vicissitudes, and other sculptures by Taylor, address consequences of climate change like ocean acidification, species decline, and habitat loss. These issues have been underrepresented topics in climate-related art (timeframe 2000–2016, Galafassi et al. 2018). Additionally, the artist also addresses the causes of degradation, such as tourism flows. In contrast to *Ice Watch,* for which mostly a negative environmental impact was assessed in the media and the generated impact report, Taylor uses a positive impact logic.

It could be argued that Taylor's works are not merely attempting local, cosmetic fixes. When considering matters around his many underwater sculptures and parks, they could be seen as developing a systems ecology, linking environmental, economic, social, cultural, and political systems. For example, in addition to the growth of marine life, the Underwater Sculpture Park influences tourism flows by redirecting divers and snorkelers to the sculpture park, and thereby it aims at preventing crowding at nearby fragile reefs (Underwater Sculpture n.d. b). Marine scientists are studying the ecosystem development. Moreover, the park has had influence on political decisions, as it 'was instrumental in the government designating the area as a Marine Protected Area. Entrance fees to the park now help fund park rangers to manage tourism and fishing quotas' (Underwater Sculpture n.d. b: para. 5). In addition, Taylor created numerous sculptures, underwater museums and sculpture parks in various locations across the entire globe (in the Atlantic, Pacific, and Indian Ocean, Mediterranean and Caribbean Sea) and entered collaborations with local communities. This co-creation with various parties situates Taylor's work in the second movement of the climate-related art framework of Galafassi et al. (2018).

Each year, over 500,000 people visit the artist's underwater museums and sculpture parks (Underwater Sculpture n.d. c). Given the popularity of the works, new questions may arise about how to manage the tourism flows to the artwork while diverting them away from fragile reefs. A comparison of the environmental impacts of Taylor's and Eliasson's public, large-scale installations could provide valuable insights, considering the scale and effort required to organize them. However, not enough data on factors contributing to CO_2, such as visitor travel and artwork transport, could be found to make this comparison.

The Desire for a Healthy Environment Is a Human Right

Taylor prefers to think about the desire for a healthy environment as a human right instead of as activism (deCaires Taylor 2021). Like *Ice Watch*, *Vicissitudes* can be experienced without necessarily knowing its meaning. The work may also be interpreted in various ways given its ongoing transformation and because spectators might understand it differently.

On the artist's website, *Vicissitudes* is described as 'a symbol of unity and resilience' (Underwater Sculpture n.d. b: para. 6). Its title 'reflects the changing conditions of the planet' (Underwater Sculpture n.d. a: The Human Face of Marine Ecologies section, para. 3). In general, Taylor's works aim to convey 'a message of regeneration and hope' when the underwater world is threatened through climate change, overfishing, tourism, and pollution (Underwater Sculpture n.d. c: para. 3). Like *Ice Watch*, *Vicissitudes* displays a circular form, viewed from above by divers or in photographs (DeLoughrey 2017). While in *Ice Watch* the circle stood for a countdown, ultimatum or an ice parliament, here the circle shape can be seen as a ring of resilience.

Others, such as the local communities, have interpreted *Vicissitudes* as an homage to the fallen slaves because of the work's proximity to the route of the slave ships from West Africa to the Americas through the

Middle Passage. The sculpture's structural supports have been taken to portray shackles (DeLoughrey 2017; van der Scheer 2021). Researchers, such as van der Scheer (2021), offer a reading of *Vicissitudes* in relation to colonialism and slavery, while referring to the threat of climate change for coral reefs. Thus, multiple interpretations of the work are possible. Taylor's sculpture seems to offer a 'plurality of meanings', 'spaces of possibility', and 'narratives of hope'—and these are seen to be important features of climate-related art (Galafassi et al. 2018: 77).

Submarine Restorationist Aesthetics

In contrast to *Ice Watch*, *Vicissitudes* and other underwater sculptures by Taylor are intended to contribute to the repair of degraded ecosystems, and thus can be seen as examples of what Demos (2016) termed restorationist aesthetics. While for Eliasson the real work of art was the space that the circle of ice invents for the people, for Taylor the real artist is the marine life: 'The marine life is much more beautiful than anything I create. It's much more intricate, much more vibrant. And that is the real artist. I'm just providing the platform for it' (deCaires Taylor 2021). Still, the human form in his sculpture suggests their significance. Even though marine life plays a large role in their formation, they are partially created by humans for humans to encounter underwater. For example, journalist Connor McGovern (2022) describes the experience of personally encountering the statues underwater guided by a diver from a dive and snorkel company: There is 'a busy wonderland of marine life' (para. 6) and 'an eerie, ghostly quality to the statues in the way they slowly fade in and out of view like apparitions' (para. 4). 'Perhaps most haunting of all, however, is Vicissitudes: a ring of life-like children, hand in hand, looking out from the circle, their eyes closed—a representation of unity in adversity, perhaps' (para. 4).

DeLoughrey (2017: 37) describes the underwater sculpture's 'submarine aesthetics', which 'are subject to an "alien" environment: transformed by salt, currents, pressure, and the rapid occupation by multispecies ecologies'. Like the ice blocks of *Ice Watch*, which were alien to the streets of major European cities and then disappeared through melting, the *Vicissitudes* sculpture initially appears strange in this underwater world, before gradually changing. In *Ice Watch*, the ice *disappears*, here marine life *grows*. Over time, it will become increasingly difficult to identify what is part of the cement sculpture and what is marine life as they merge into one. The sculpture will become indistinguishable from the reefs.

A divide into environment, the ocean, the marine life on the one hand (nature), and sculpture (which can be seen as a symbol of culture), on the other, will no longer apply. Nature becomes inseparable from culture, and *naturecultures* emerge.[5] The sculpture (and perhaps a kind of human centring) will be transformed by the underwater world into something more-than human (van der Scheer 2021).

The artist situated the sculpture at shallow depths enabling a certain public to view them, like snorkelers and glass boat passengers. However, more people will likely encounter the sculpture represented in photography. Thus, while being an experiential work, the representation through images plays a significant role similar to *Ice Watch*.

Conclusion and Discussion

This chapter asked: *How do visual climate-related artworks employ aesthetics and engage with politics and structural causes of the climate crisis?* First, some works focus on both the structural causes and the consequences by *representing* systems (particularly Biemann) and *developing* a systems ecology (especially Taylor when the cultural, economic, political and environmental systems and relations around his many sculptures are considered). Others primarily visualize the impacts of the climate crisis (Eliasson and Rosing), but also to some extent merge impacts with the causes (Burko). Impact in these artworks has two important meanings: the climate effects they address and the footprint of the work itself. While transport of works and visitor travel associated with large-scale installations like *Ice Watch* are discussed as major contributors to CO_2e in the visual arts (Bottrill et al. 2021), other aspects are still neglected, including the digital impact of online art presentations. More transparency and research are also needed across all art forms and regarding recent developments such as AI art, which can involve substantial server use and which raises ethical questions. In addition, all arts practices, and particularly large cultural institutions and fairs, whether climate-focused or not, should consider their environmental and social responsibilities. The necessity to include this so-called 'sustainability in the arts' aspect (a term used and described by Power 2021) in the discussion of art has been expressed by numerous scholars and practitioners in recent years.

[5] Donna Haraway, after Bruno Latour, has coined the term *naturecultures* (Neimanis et al. 2015).

Second, particularly Biemann's work can be seen as part of a political ecology approach. She pays attention to power relations and people, and focuses particularly on the politics of environmental justice. Biemann's and Burko's work are clearly addressing climate change through their climate-related videographic storyline and representation. In contrast, the work of Eliasson and Rosing, as well as Taylor, can be experienced without necessarily knowing the underlying climate messages—namely, that arctic ice is melting and urgent climate action is needed (Eliasson and Rosing), and that the marine world is threatened but regeneration is possible (Taylor). When considering the timing, location, and explanations about the works, their activist function becomes clearer. Especially *Ice Watch* coinciding with COP and IPCC meetings can be seen as climate activism. Although the four artworks do not contain explicit activist calls, activism is openly voiced on the artists' websites (Burko, Eliasson, Taylor), through public engagement (Burko), and through their affiliation with collectives engaged in activist work (Biemann). While their artistic and activist activities may overlap, the borders are not always clear-cut or clearly distinguishable.

Third, the aesthetic emphasis of the artworks spans a diverse range, each employing aspects of the sublime to engage with the climate crisis: the industrial sublime in Biemann's video essay, Burko's referential abstract sublime, the dark sublime of melting arctic ice in *Ice Watch*, and the natural sublime in and around Taylor's underwater sculpture. The beauty of nature can be seen as the basis of all works, even if to different degrees. Furthermore, an aesthetics of sustainability (Kagan 2011), in the sense of a sensibility to complexity and sensibility to connected patterns, can be particularly experienced in Biemann's representational work.

Different audiences will engage with these works in their own way, influenced by their individual presuppositions and interests. Some may prefer the referential abstract sublime aesthetics of Diane Burko. Others will be interested in Ursula Biemann's research-oriented eco-realism. Yet others participate in the experiential works of Eliasson and Rosing or Taylor, or view their representations online. The artwork analysis reveals that the distinction between representational and experiential works blur as experiential works like those of Eliasson and Taylor are documented through photography and video and shared across news outlets and social media platforms.

Acknowledgements This research was made possible through a grant from the Dutch Research Council (NWO) (PGW.19.010/8366). I would also like to thank all artists for providing the artwork images for this chapter, and my PhD co-supervisor Prof. dr. Katja Kwastek and the editors of the book, Daniel Binns and Rebecca Najdowski, for giving valuable feedback about this chapter.

References

Alexander, Victoria D. (2021), *Sociology of the Arts: Exploring Fine and Popular Forms*, Hoboken, NJ: John Wiley & Sons.

Biemann, Ursula (2016), 'Deep weather', *Geo Humanities*, 2:2, pp. 373–376.

Bottrill, Catherine, Tickell, Alison, Buckley, Claire, Melitsko Thornton, Graciela and Badiali, Chiara (2021), THE ART OF ZERO—An indicative carbon footprint of global visual arts and the transition to net zero, https://juliesbicycle.com/wp-content/uploads/2022/01/ARTOFZEROv2.pdf. Accessed 11 November 2023.

Burko, Diane (2021), online interview with U. Hahn, 7 May.

Burko, Diane, Garrard, Mary D. and Broude, Norma (2021), *Diane Burko: Seeing Climate Change, 2002–2021*, exhibition catalogue, American University Museum at the Katzen Arts Center, Washington, 28 August–12 December.

deCaires Taylor, Jason (2021), online interview with U. Hahn, 22 November.

DeLoughrey, Elizabeth (2017), 'Submarine futures of the Anthropocene', *Comparative Literature*, 69:1, pp. 32–44.

Demos, T. J., Scott, Emily E. and Banerjee, Subhankar (eds) (2021), *The Routledge Companion to Contemporary Art, Visual Culture, and Climate Change*, New York: Routledge.

Demos, T.J. (2016), *Decolonizing Nature. Contemporary Art and the Politics of Ecology*, Berlin: Sternberg Press.

Galafassi, Diego, Kagan, Sacha, Milkoreit, Manjana, Heras, María, Bilodeau, Chantal, Bourke, Sadhbh J., Merrie, Andrew, Guerrero, Leonie, Pétursdóttir, Guðrún and Tàbara, Joan D. (2018), '"Raising the temperature": the arts on a warming planet', *Current Opinion in Environmental Sustainability*, 31, pp. 71–79.

Geobodies Official Website (n.d.), 'Deep weather', Geobodies, https://geobodies.org/art-and-videos/deep-weather/. Accessed 11 November 2023.

Ginn, Franklin, Bastian, Michelle, Farrier, David and Kidwell, Jeremy (2018), 'Introduction: unexpected encounters with deep time', Environmental Humanities, 10:1, pp. 213–225.

Hornby, Louise (2017), 'Appropriating the weather: Olafur Eliasson and climate control', *Environmental Humanities,* 9:1, pp. 60–83.

Ice Watch London Official Website (n.d.), 'Ice Watch', https://icewatch.london/. Accessed 3 August 2023.

Jackson, Michael (2015), 'Glaciers and climate change: narratives of ruined futures', *Wiley Interdisciplinary Reviews: Climate Change*, 6:5, pp. 479–492.

Julie's Bicycle (2019), 'Ice Watch London', Julie's Bicycle, 8 November, https://juliesbicycle.com/news-opinion/ice-watch-london-2018/. Accessed 3 August 2023.

Kagan, Sacha (2011), *Art and Sustainability: Connecting Patterns for a Culture of Complexity*, Bielefeld: Transcript Verlag.

McGovern, Connor (2022), 'Exploring Grenada's underwater sculpture park', National Geographic, 20 May, https://web.archive.org/web/20220521012037/https://www.nationalgeographic.co.uk/travel/2022/05/exploring-grenadas-underwater-sculpture-park. Accessed 3 August 2023.

Morton, Timothy (2013), *Hyperobjects: Philosophy and Ecology after the End of the World*. Minneapolis: University of Minnesota Press.

Neimanis, Astrida, Åsberg, Cecilia and Hedrén, Johan (2015), 'Four problems, four directions for Environmental Humanities', *Ethics & the Environment*, 20:1, pp. 67–97.

Olafur Eliasson Official Website (n.d. a). 'Ice Watch, 2014', https://olafureliasson.net/artwork/ice-watch-2014/. Accessed 3 August 2023.

Olafur Eliasson Official Website (n.d. b). 'Ice Watch, 2015',https://olafureliasson.net/publication/ice-watch-2015/. Accessed 26 July 2023.

Ouellet, Maryse (2020), 'Within aesthetic distance: artistic critique from activism to eco-realism', *Konsthistorisk Tidskrift (Journal of Art History)*, 89:2, pp. 126–149.

Power, Kate (2021), '"Sustainability" and the performing arts: discourse analytic evidence from Australia', *Poetics*, 89.

RADIUS (2022), *Emotions are Oceans*, exhibition catalogue, RADIUS Center for Contemporary Art and Ecology, Delft, 8 May–26 June.

Roosen, Liselotte J., Klöckner, Christian A. and Swim, Janet K. (2017), 'Visual art as a way to communicate climate change: a psychological perspective on climate change-related art', *World Art*, 8:1, pp. 85–110.

Rosenblum, Robert (1961), 'The abstract sublime', *Art News*, 59:10, pp. 38–41.

Sommer, Laura K. and Klöckner, Christian A. (2019), 'Does activist art have the capacity to raise awareness in audiences? A study on climate change art at the ArtCOP21 event in Paris', *Psychology of Aesthetics, Creativity, and the Arts*, 15:1, pp. 60–75.

UN Climate Change (2015). 'Arctic ice art displayed in Paris', YouTube, 4 December, https://www.youtube.com/watch?v=Tpe4o9_n8AM. Accessed 6 August 2023.

Underground Channel (2014). 'Ice Watch, 2014', SOE.TV, 30 October, https://www.soe.tv/video/ice-watch-2014. Accessed 6 August 2023.

Underwater Sculpture Official Website (n.d. a), 'Overview'. https://www.underwatersculpture.com/about/overview/. Accessed 10 November 2023.
Underwater Sculpture Official Website (n.d. b), 'Molinere Underwater Sculpture Park', https://www.underwatersculpture.com/projects/molinere-underwater-sculpture-park/. Accessed 6 August 2023.
Underwater Sculpture Official Website (n.d. c), 'Home', https://underwatersculpture.com/. Accessed 5 May 2023.
van der Scheer, Inez B. (2021), 'Everything is alive: Jason deCaires Taylor's Vicissitudes', in T.J. Demos, E.E. Scott and S. Banerjee (eds), *The Routledge Companion to Contemporary Art, Visual Culture, and Climate Change*, New York: Routledge, pp. 376–383.
World of Matter Official Website (n.d.). 'About the project', https://www.worldofmatter.net/about-project. Accessed 7 August 2023.
Yusoff, Kathryn and Gabrys, Jennifer (2011), 'Climate change and the imagination', *Wiley Interdisciplinary Reviews: Climate Change*, 2:4, pp. 516–534.
Zarin, Cynthia (2015), 'The Artist Who Is Bringing Icebergs to Paris', *The New Yorker*, 5 December, https://www.newyorker.com/culture/culture-desk/the-artist-who-is-bringing-icebergs-to-paris. Accessed 3 August 2023.

CHAPTER 7

Networked Photography and 'Image That Comes'

Ana Peraica

Introduction

With the enlargement of online image databases and the merging of public and personal archives, our knowledge about history grows, producing what Martin Lister names a 'collective memory of the internet' (Lister 2014: 126). Some of these images—like family, tourist, and memorial pictures—were never meant to be public. These are sometimes called 'snapshot photography' (de Duve 1978: 113; Batchen 2008: 12; Berger 2011: 175), 'domestic photography' (Spence and Holland 1991), 'vernacular photography' (Batchen 2000: 56), 'ubiquitous photography' (Hand 2012), or 'banal imaging' (Jurgenson 2019: 145–147). Today most of them are becoming 'networked images' (Niederer 2018), or images that are uploaded on the net, and distributed online. Some of these images, especially from travel, capture the world as seen by everyday people, rather than by the tourist industry or corporations. Such is the

A. Peraica (✉)
Atelier Peraica, Split, Croatia
e-mail: anaperaica@gmail.com

D. Binns and R. Najdowski (eds.), *Confronting the Climate Crisis*, Palgrave Studies in Media and Environmental Communication,
https://doi.org/10.1007/978-3-031-89606-4_7

case with the image of marketing specialist MD Jawadur Rahman, who uploads photographs to the Pexels platform that depict his own homeland as well as his personal travels. Figure 7.1 is one of Rahman's photographs, depicting an old man looking up at the camera; a birds-eye perspective focuses the view on the subject. The photograph's depth of field is moderate; though slightly blurry, the background shows rubbish floating in the river. Although wearing worn and wearing clothing, the man in the foreground is clean, while the background shows the pollution, grit, and grime: the world of the instant.

Most private images, as well as professional ones, deny the problem of ecology by aesthetic framing of the image, omitting the 'ugly', contaminated parts of nature. Some still do document it, but without giving much attention to it. The reason is quite banal; being highly private, as Rubinstein and Sluis warn (2008), snapshot photographs are not always influenced by dominant professional aesthetics (Nickel 1998). Because these amateur images are not framed according to advertising's conventions, nor do they rigorously adhere to scientific or documentary 'truth', they offer an uncensored, perhaps more objective, reality.

Fig. 7.1 MD Jawadur Rahman, *Elderly ethnic man near dirty river*, 2020, Photograph, 6000 × 4000px. Location unknown. Copyright-free. https://www.pexels.com/photo/elderly-ethnic-man-near-dirty-river-3719494/

The Anthropocene

The Anthropocene discourse is concerned with the relationship between a past in which mistakes were made, and a living present in which the effects of those mistakes are becoming visible. But any Anthropocentric perspective also considers the unpredictability of any future that may result from this relationship. It shifts our questions from the passive reception of ecology to the active analysis of our own personal trails through the landscape. Although meant to describe geological temporalities, the term Anthropocene has also entered the lexicon of many academic disciplines including the humanities. As Anthropocene debate revolves around the interplay between a past marked by yet-invisible transformations and a living present in which those transformations are manifesting, it presents speculation as to an uncertain future. The domain of these analyses is known by various names, with environmental humanities being the most common and widespread.

Environmental humanities are fundamentally an interpretive discipline that connects social and humanistic studies to the ecological impact that humans have made on their own environment. This interdisciplinary domain incorporates the Anthropocene paradigm derived from the natural sciences, specifically geology, as well as botany, zoology, and biology, into the core area of the humanities. The aforementioned transition influenced numerous academic disciplines, which have since realigned themselves to investigate ecological concerns, such as history, anthropological or contemporary archaeological studies, and art history. The latter is present in the emergence of a more comprehensive Anthropocene art history and the subsequent theoretical insights it has produced, including eco-critical art history.[1] Furthermore, numerous case studies apply to image studies, with reverberations back to scientific disciplines (Weintraub 2018; Kastner and Wallis 2010; Cheetham 2018), implying that they do not remain isolated but rather integrated academic cases. In addition to researching the artistic and common visual representation of environmental changes, these approaches seek to understand why

[1] Anthropocene art history (Demos 2013; Davis and Turpin 2015; Demos et al. 2016, 2017; Corbett and Clark 2017; Hedin and Gremaud 2018; Reiss et al. 2018), is distinguished from approaches known as eco-critical art history (Braddock 2009, Braddock 2014, Clark 2015, Coughlin and Gephart 2019).

some of these representations, which call for immediate action, received insufficient response or critical reception.

Environmental Photography

The Anthropocene is said to have been formed around the impact of a steam engine, a great polluter whose massive employment fueled the series of inventions known as the Industrial Revolution. Soon after its implementation, a slew of new inventions began, some of which directly incorporated new steam machines as a drive for larger ones, such as steamships and steam locomotives, while others took a new direction in enhancing travel and bridging distances. The transportation industry connected space and air with telecommunications. Among other things, this development brought with it the inventions of automobiles and telephones, as well as electricity. The world changed. Some of these changes did not appear to be significant at first, but over time they accumulated. To halt the world that was changing at great speed, another tool was invented: photographic recording, registered in 1839 (the year then commonly seen as its official birthdate).[2]

As the world's landscape changed with the transportation industry, cities were expanded to include travel and industrial zones, and roads were built to accommodate new vehicles. Not to mention how much nature changed with the extraction of coal and later on oils. A photographic camera was followed by recording new 'achievements'. Nineteenth-century photography captured many images of polluted cities, wholesale environmental changes caused by massive urbanization and deforestation, and geological material exploitation, but at first such images were not seen as critical. Images of chimneys well illustrate this. Today, we view nineteenth-century chimneys as the quintessential symbol of the Anthropocene. They were already present in the oldest photographic, or more precisely, heliographic, image—Joseph Nicéphore Niepce's *View from the*

[2] Few photographers, experimenting with systems of fixation of the image autonomously, arrived at the process of fixation we today know as photography. Among them were Hippolyte Bayard, Nicephore Niepce, and Hercules Florence, while another process was pending by William Henry Fox Talbot, yet the invention announced to the French Academia of Science in 1839 was the one by Jean Lous Mande Daguerre, featuring a direct positive image.

Window at Le Gras from 1826—but here they are those of the bakery.[3] Soon after, images of chimneys that were crowded in small Victorian cities followed. Some cities had a remarkable chimney scape; for example, only Manchester had approximately five hundred of them by 1843 (Bonneuil and Fressoz 2016), and these chimneys were recorded at various stages of their construction. However, these photographs were not made as critical comments but rather served to document the city's development, as Steven Mosley elaborates in Chimney of the World (2001), writing on pollution in this leading center of the Victorian textile industry. Although pollution has already had fatal consequences for health, it has not been recognized as harmful by the majority; rather, it has been seen as a mere epiphenomenon of industrial success. Photography became a literary 'silent witness' to human environmental damage, documenting it. The paradox of these images is that, while images of pollution have existed since the invention of the stabilized photographic procedure, awareness of it has been delayed. As environmental knowledge was postponed, photographic evidence of the lost world was also created, complicating the relationship with nature. This was due to the continuation of landscape genre aesthetics, which was inherited from painting and continued in photography. To understand these developments, we must take a few steps back.

The Curse of Picturesque Image

One of the reasons why critical awareness of pollution did not arrive in time while actually going on and being recorded can be found in the importance of the landscape genre in painting of the nineteenth century, closely associated with the idea of picturesque. Picturesque refers to a romantic vision of land, inherent to the logic of the picture itself; its decorative role is important for the living environment in which it had to decorate, appropriating the external world to halt its terrifying changes.

[3] Displaying, as Barbara Brown writes;

'The view, made from an upper, rear window of the Niépce family home in Burgundy, represents (from left to right): the upper loft (or, so-called 'pigeon-house') of the family house; a pear tree with a patch of sky showing through an opening in the branches; the slanting roof of the barn, with the long roof & low chimney of the bake house behind it; and, on the right, another wing of the family house' (Brown 2002).

Rosalind Krauss, a photography theorist, explains the relationship between landscape and environmental changes (Krauss 1982). She notes that the evolution of the landscape genre is inextricably linked to human changes introduced around and on land. They are the result of humans' perceptions of their surroundings as well as modern awareness. Images of the land were liberated from the human figure and began to represent only nature in the fifteenth century (Gombrich 1964), but at first, they were only iconographic or symbolic. Yet, it took until the seventeenth century for environmental representation to become a common theme. However, the reason for its occurrence was more utilitarian than an appreciation for the beauty of the environment. It was a documentation and report on new world discoveries, bringing information on new possible settlements in the era of completing the 'world picture'.[4] Tied to the contestation of territory, despite their beauty, these pictures were all but innocent. Documenting various stages of initial discovery, conquering, replacing, or completely erasing indigenous nations and autochthone species with new land cultivation, such images frequently speak to the political economies of the times in which they were created. Organized colonialism and imperialism quickly followed the discovery of new continents, bringing with them violence through land expropriation, confiscation, and even the destruction of combat with the original owners (Mitchell 2002). More specifically, the landscape initially represented the colonization of other spaces (Bermingham 1987; Mitchell 2002), as well as agrarian reform logic and land management. Many authors thus attribute the origins of the landscape genre to political land management. So, Mitchell referred to the genre as 'the dreamworld of imperialism' (Mitchell 2002), while Ann Bermingham defined landscape as an ideology in and of itself (Bermingham 1987).

Since the seventeenth and nineteenth centuries, when society completely transitioned from feudalism to national capitalism, the landscape has displayed various forms of land ownership (Cosgrove 1988). Images at the time were used to describe topographical features of the countryside, often documenting the size of a property. In relation to

[4] New visual fashion was so popular that geographer Edward Casey referred to this era, evoking Heidegger's idea of 'world picture' as 'the age of world picture' (Casey 2001), defining the moment from which humans have started to perceive the world as a genuine place.

these images, geographer Cosgrove wrote that the landscape served to represent property relationships:

> with the hegemonic establishment of urban industrial capitalism and the bourgeois culture of property, landscape lost its artistic and moral force and became a residual in cultural production (Cosgrove 1988, 64).[5]

However, with the development of capitalism, the landscape also began to change dramatically, and it was documented far more often. This is when photography appeared on stage. Initially, the photographic landscape genre cannot be said to directly address the need for documentation of damaged landscapes. It rather continued to fulfill the need for a decorative, picturesque image on the wall, representing a place to settle down or proprietary land to be admired. Yet while some images appeared beautiful back in that day, today they reverberate completely different aesthetics, like those Manchester chimneys. Time widened the gap between idealized nature, marketed as a desirable destination, and today's polluted land.

Regressive and Progressive Landscape

Such a split was already visible in painted landscapes, so WJT Mitchell, in his *Landscape and Power*, recognized a 'double structure' of landscapes: regressive or nostalgic and progressive or colonial (Mitchell 2002, 17), or landscape as decoration and landscape as documentation. Although Mitchell did not write specifically about ecology, his distinction between regressive and progressive landscapes can now be used in the environmental humanities to approach the distinction between an idealized image of clean and a critical image of polluted nature. Similarly to Mitchell, Jim Elkins in his *Landscape Theory* (2008) referred to Stephen Daniels' concept of a 'duplicity of the landscape' (1988) contrasting subjective experience from a rational one, the first being a romantic concept while the second one belonging to the political or economic authority, such as land ownership (Elkins 2008, 27). One is our perception of the landscape, and the other is its actual condition. Elkins' distinction can also be used to

[5] Cosgrove recognized in such a landscape a proper sign of early capitalism's hegemony, as peasants were not represented, whereas their exploiters were portrayed more frequently (Cosgrove 1988). Even when depicted, the rural poor were merely ornaments to the image, appearing as urban subspecies (Barrell 1980).

assess the difference between propaganda and reality. Similarly, archaeological theorist Michael Shanks (2016) examined two types of landscapes: picturesque views and visual evidence, highlighting the stark contrast between artistic and scientific perspectives on the land. Such distinction is useful to analyze the differences in the use and function of landscapes as decoration and knowledge, and to differentiate their specialist fields.

Production of Anthropocentrism and its Critique

The proactive role of art production and critically focused exhibitions in addressing concerns about human accountability for the destruction of life on the planet and the variety of its inhabitants has contributed to the growing attention and diverse range of reactions in culture and the arts.[6] However, as demonstrated by a number of authors, artists' warnings from the nineteenth century to the present, particularly those depicting contaminated landscapes, have frequently been ignored. Paradoxically, as some researchers noted, the ecocritical discourse in the arts has existed since land art in the 1960s and has had a consistent presence at major shows such as the Venice Biennial, but its reception, rather than production, has been somewhat marginal until the current climate awareness has taken center stage, as the addressed public was already aware of these problems. Similar was the destiny of the earliest known show to critically discuss the theme of human-nature relationships. This appears to be the exhibition *L'homme contre nature*, which was presented at the Musee d'Historie in Paris in 1955, the same year that Edward Steichen established the canonical show *Family of Man*, centering post-war anthropocentrism around the idea of the universality of human experience or even human condition, which remains constant despite cultural differences, and it emerged as a show that echoed the idea of post-WWII reconciliation.

The concept of Steichen's show was back then heavily criticized for its generalizations (Barthes, 1957).[7] But, today it could be criticized even further for consolidation of the hard version of anthropocentrism

[6] To refer to two notable examples from the past: *Earth: Art of Changing World*, The Royal Academy of Arts, 3 December 2009, and *Unfold*, The Glass Curtain Gallery, 26 November 2010–22 January 2011.

[7] Nonetheless, it has received a lot of attention in terms of critical theory, and it is still regarded as one of the most important exhibitions of all time (Batchen 1997).

in general; and further as a promoter of globalization with its disastrous impact on the environment. Despite Steichen's large personal interest in biology, as he was himself cultivating and exhibiting hybridized plants, this show had little to do with nature.[8] In contrast to Family of Man, the exhibition *Man Against Nature*, received lukewarm reviews and was poorly documented. What is documented is an amazing worry of mid-twentieth-century intellectuals regarding the loss of species due to human action. An issue of UNESCO's *Courier* from January 1958, featuring the same title as the exhibition Man Against Nature, proves this.[9]

Despite these reception and reverberation differences, Steichen's show and UNESCO's issue had diametrically opposed points of view, with one criticizing the system established by the other. They also had one crucial similarity. They both used photography as their medium of documentation and communication. Used to promote two completely different ideas, anthropocentrism and its critique, the position of photography can be found paradoxical; it is both the producer of environmental degradation and its documentarist. Photography, a perspectival apparatus of the Anthropos, is a perfect shadow of the Anthropocene, following it nearly since its beginnings, for which reason I named it—the Photographocene (Peraica 2021).

Greenwashing Landscapes and Anthropocene Records

Finally, the closest distinction has been made by photography historian and theorist Liz Wells in her recent volume *Land Matters* (2022). Wells

[8] He cultivated primarily delphiniums, poppies, and sunflowers, which he exhibited at MoMA (1936), curated by Alfred Barr. It is also notable that George Gessert, continuing Steichen's work, created thousands of irises and installed hundreds of photographs in the series Hybrids (1991).

[9] This print issue cautions against a number of factors that contribute to the extinction of species, including hunting tourism in Africa, so-called Big Game, cases in which animals are killed for destroying human objects such as fishing nets, or even because of superstition. An inventory of animals that were already endangered at the time is provided in the article 'Fossils of Tomorrow' by Marguerite Caram and J.J. Peter: the lion of Asia, the mountain zebra, the right whale, the one-horned rhinoceros, the Addo elephant, the wild bison, the Caribbean monk, lemurs, and the koala, among others. An alternative article lists some extinct species, including the dodo, great auk, quagga, and rhytina. Ultimately, the matter illustrates the endeavors of numerous nations to increase public consciousness regarding the extinction of species through initiatives such as the establishment of expansive nature parks and the printing of postage stamps.

emphasized developments from picturesque to sublime landscape, the first realizing itself in photographic souvenir, whereas 'By contrast with the beautiful and picturesque, the sublime is associated with awe, danger, and pain, with places where accidents happen, where things run beyond human control, where nature is untamable' (Wells 2022, 48). What all four distinctions between two types of images have in common is that they define how human politics changes and influences the landscape. Aestheticized images can mask the presence of critical ones due to their quantity and redundancy, especially when aesthetics serves as demagogy or for entirely economic reasons. Institutional landscapes, on the other hand, can advocate for and propagate specific conditions.

Today, we can clearly see the difference between the two types of landscape photographs: those that beautify and romanticize the environment and those that provide documentary value, as it is enlarged. Unfortunately, because of the marketing machinery, particularly the travel industry, that now stands behind such images, the attractive image of clean nature prevails. Flora Mary Barlett refers to the continuity of the cult of 'pristine' or an untouched nature, which comes from the Romantic tradition of the unorganized or wild nature as untouched, with highly edited pictures in tourist brochures showing lifestyles (Barlett 2023, 29). These images are also part of the so-called 'greening' or greenwashing propaganda campaign; showing how the nature revitalizes by some strange miracle. In contrast to such seductive images, there are also ones that we rarely refer to as landscapes: images in which the same nature is hostile, contagious, or poisonous. Polluted landscapes, commonly not having a decorative function. Images of damage are not only rarer, as inaesthetic, but also appear to be more difficult to record than beautiful landscapes.

Visual theorist Nicolas Mirzoeff takes a step further in analyzing the possibilities of images of such an impact at all, showing that a part of this impossibility of seeing the environmental damage is also a problem inscribed in the media. He is claiming that representations of the Anthropocene really work against our attempts to understand our own future because they coproduce human-caused damage and serve to obscure the true problem of environmental degradation rather than showing it as it is. We cannot understand the human impact on the world in its fullest, and most likely final, sense since these images are so ingrained in our imagery that they are indistinguishable or have yet to be fully realized. Thus, we cannot understand the world without us not only because it has not yet arrived, but also, we cannot exclude ourselves. Besides, most images produced contribute to the Anthropocene by perpetuating the exploitative capitalist status quo: encouraging us to actively participate in environmental degradation by traveling long distances, purchasing plastic products, and using more and more computing power. Very few of the photographs produced daily serve to raise awareness of our own responsibility for potential personal carbon footprints or other environmental damage. What we see are mostly the results we've come to expect, and they appear to happen regardless of our actions. Mirzoeff thus concludes, '[t]he Anthropocene is so built into our senses that it determines our perceptions' (Mirzoeff 2014: 223). Mirzoeff refers to this period as 'the era of countervisuality' (Mirzoeff: 223).[10] Irmgard Emmelhainz acknowledges the reduction of Anthropocene imagery on apocalyptic phantasy, typically presented as 'fixable catastrophe' (Emmelhainz 2015: 1), concluding that, as a result, the true images of the Anthropocene remain missing (Emmelhainz: 9). They are yet to come. But then the question is—would we be there to see them. And are there images that record the environmental degradation that we actually do not see.

Image that "Comes"

While we are overwhelmed with images of pure nature in print marketing, TV commercials, and internet pop-ups, cities are growing, urbanization and industrialization are speeding up, deforestation escalates, and the sea level is rising. To visualize these processes as proof that the environment has altered, we often need two images: one before and one after the change took place, as in the case of cities emptied by COVID-19. Setting

[10] This is often related to the economic system of capitalism, the reason being that the visualization of the Anthropocene is directed by the capital. The promise of a future (and salvation) in products is necessary for this capital to continue ruling the planet. However, what we fail to see, or fear to imagine, is our own vulnerability in this world. According to Mira J. Hird:

> 'What characterizes an Anthropocene aesthetics is a heightened anxious slippage between visualizations of the conquest of nature and the realization of our vulnerability *as* nature' (Hird in Colebrook and Weinstein, 2017, orig. ital.).

these images next to each other makes the transformation clear. Changes are then compared and monitored using a few visual comparison models that differ in their purposes. Before and after sequences used to compare historical and current changes (Bear and Albers 2017) can be organized more strategically as repeat and time-lapse photography. Repeat photography uses an archival image to analyze changes that occurred (Webb 2010), so it intentionally repeats the same spot and angle, as well as lenses and camera settings if possible. Time-lapse photography situates the camera on-site to automatically take sequential images in order to monitor changes (Buckley et al. 2017). One model is used to capture differences that occurred over a longer time span, the other to notice slightly faster changes such as sea levels and ice melting.

Perhaps the most visible among these rhetorical setups is made by comparing images of melting glaciers, which have become symbols of humanity's destruction of Earth's balance (Garrard and Carey 2017). Older images taken in comparison were not intentionally created to document active defrosting but were photographs taken by expeditions in the nineteenth or early twentieth centuries. So for example, Chrystel Lebas in 2011 rephotographed early twentieth-century glass plates of landscape scenes of Scotland, Norfolk, and Devonlandscape by British botanist and ecologist Sir Edward James Salisbury (1886–1978), kept in the Natural History Museum London.[11] Tyrone Martinsson created a re-photographing project (2016) using images captured by explorer and photographer Nils Strindberg, who died during a balloon expedition in the Arctic with researchers Andrée and Knut Frænkel.[12] Five rolls of Strindberg's undeveloped film, one still in the camera, were discovered with their bodies in the Kvitøya area. While the majority of the material suffered from temperature changes and being soaked in water, a portion of these negatives were in good enough condition to be developed and used for active comparison of landscapes.[13] Glaciers are actively monitored by the USGS's *Repeat Photography Project*, which raises awareness

[11] Chrystel Lebas *Revisiting* (2014) http://www.chrystellebas.com/Re-visiting/re-visiting.htm.

[12] Besides, Martinsson also uses films on the expedition to Artic by Benjamin Lee Smith in 1873, as well as some lithographs from 1865 (Wells 2022, 86).

[13] They were published as *Med Örnen mot Polen* (1930) and *The Andrée diaries* (1931);

on climate change using this simple comparable method.[14] In addition to monitoring glaciers, scientists create new images to capture processes in real time. So, for example, photographer James Balog founded Extreme Ice Survey (EIS), a photographic team actively surveying glacier melting in many locations, including Greenland, Iceland, the Alps, Alaska, the Rocky Mountains, Bolivia, and Nepal.[15] Photographs were published as a monograph, Portraits of Vanishing Glaciers (2012). However, Balog's photographs differ from those of Lebas and Martinsson in that they are more focused on the aesthetics of phenomena and occasionally their attractiveness.

Aesthetic of photography is also dominant in photographic images compiled by Patricia Macdonald's illustrated book *Surveying the Anthropocene: Environment and Photography Now* (2022), which examines recent artistic projects that document environmental degradation and its consequences. Perhaps the most well-known are Edward Burtynsky's *The Anthropocene Project* series.[16] These images depict natural landscapes that are visibly changing as a result of deforestation, agricultural cultivation, oil fields, urbanization, landfills, chemical pollution, light pollution, various types of migration, nuclear tests and bombs, and oil spills caused by human population growth at the expense of other species that are drowning, suffocating, burning, or starving. Burtynsky shot them in Ural in Russia, India, Nigeria, Chile, and China, but also in Berlin, illustrating the massive impact on the landscape in both distant and near spaces. Many of these images are taken from unusual angles, primarily aerial, or in locations we have never visited. They are not part of many people's daily lives, and if they are, then they are usually recorded in poorer countries where toxicity and weather disruptions are visible on a large scale.

Despite being visually appealing and adhering to autonomous aesthetic rules of visuality, such as the golden ratio and the photographic rule of thirds or fifths in composition, Burtynsky's images are beautiful but emotionally devastating. Yet, precisely for being visually appealing and depicting damage to the Earth's crust, these images also evoke thoughts about the prevalence of aesthetics over ethics and politics, which is

[14] *Repeat Photography. Project* https://www.usgs.gov/centers/norock/science/repeat-photography-project.

[15] *Extreme Ice Survey*. http://extremeicesurvey.org/

[16] Others to mention are Lucas Foglia, Subhankar Banerjee, etc.

inherent in the nature of photography. The fact that we may actually like images depicting the loss of natural habitats distinguishes the image's non-decisive point; the camera does not answer the question of whether images are artistic or scientific, staged or not; only their context does. Moving these images from galleries to newspapers, one may be astonished to discover how the idea of sublime and the anesthetization of extinction may challenge the one of warning and an urge to action. To amplify this urge, Macdonald's edited book, in addition to Burtynsky's images, also refers to those created by scientists and citizens, as well as images that are not intended to depict the theme of the Anthropocene. Despite the large number and variety of perspectives, the reader is unable to grasp the complexities of environmental degradation, but this is not due to artistic or editorial reasons. The environmental damage is incomprehensible and unphotographable in and of itself. Besides it is also published in limited context which usually is already ecologically aware (as art public is, for example).

While photography clearly captures specific instances of human-caused damage to the environment, it cannot capture climate change in its entirety due to its enormous scale and complexity, which does not correlate to the scale and complexity of the photographic medium (Fisher 2013), as well as various pollutions, such as nuclear ones. Oftentimes, the most dangerous threats to specific environments, such as toxicity, radiation, and viral contamination, are completely invisible and therefore cannot be captured using standard photographic equipment. Only its effects are observable. These phenomena are, thus, only recorded as symptoms. So instead of the concept of extreme weather, what we see in these images are individual and site-specific cases of hurricanes, tsunamis, fires, floods, and droughts. But these are only the consequences, not the reasons of environmental changes. A similar situation exists with contamination, where the only visible effects are the effects of contaminated clouds or dirty water, not the contamination itself. Images of electronic waste landfills and garbage dumps stand for the whole system of toxic waste disposal and the economy of garbage. But while we see these images as consequences of previous actions, we cannot see the consequences of our own actions—that includes the one of photographing.

But, here and there, we are already faced with bare images stripped of any poetic qualities, confronting us ruthlessly with the future of life. Images of melting glaciers, rainforests destroyed by fires, and the spread of toxic algae foreshadow the future image of the same places and announce the continuation of these processes in the future. We do not see the fullness of these processes, only the initial start of the process.[17] Such images of something that is only announced and has not yet occurred in its entirety are inherently contradictory to the common photographic image, which always records only the present for the past and stores it for future memory. For example, the Greenpeace organization released a 1997 image of a crack in an Antarctic ice shelf, which significantly increased awareness. Similarly to prisoners sentenced to death awaiting trial, we can see landscapes that have been polluted and contaminated by man, as well as ones that have not yet been polluted but will suffer the same fate of a global ecological crash, bringing death as an invisible but necessary future. We can imagine what Kember refers to as 'environments that could be' (Kember 2017, 348), or images of the future that are actually missing, depicting something that has not yet happened in the fullest sense, which is only announced and 'yet to come' (Emmelhainz 2015, 250). They belong to what Stengers calls 'future perfect continuous tense' (Stengers 2015, 139). Yet, while the process is certain, the future might not be. But unfortunately, such changes had already occurred at some places, as is the case with an old man in the image.

Invisible Garbage

Living in the culture of disposed goods, a so-called 'throwaway culture', centered around a new paradigm of the product as temporary (Hell and Schönle 2010), produces a problem of agglomerating waste in packages and wraps, especially since the invention of plastic in the 1950s. In a lecture in the framework of the event *Critical Zones* by ZKM in Karlsruhe, a geologist, Jan Zalasiewicz (2008), analyzed the average meter square of a soil.[18] Contrary to some beliefs that a land is different at various places, what Zalasiewicz proved is that economic globalization also influenced the globalization of the toxicity of individual land. In the average meter square, in addition to old materials such as concrete, plastic, aluminum, iron, and steel, as well as carbon dioxide, coal, and burned oil, there are also an amount of objects such as cigarettes, computer chips, living

[17] This can only be done with visualization tools, which can be implemented to see the future, using photographs that document the past.

[18] Jan Zalasiewicz, *Visualizing Anthropocene*. https://www.youtube.com/watch?v=zphinkipB9A&t=143s

matter, shrubs, and the remains of supermarket chickens. As a matter of fact, there is hardly any place without microplastic in the soil or a 'clean soil'. Moreover, synthetic materials appear to be more durable than many geological structures, including stones. Tragically enough, the life span of plastic is so long that it will surely outlive even the ancient monuments and life finding its place inside of it. Instead of disintegrating, plastic merges with other materials, forming a strange mass known as *plastigomerate* (Robertson 2016). Although seen as being an omnipresent structure, this mass of plasticized materials, even more densely structured than by plastic itself, is appearing at global dumping yards and landfills, including India and China, among other countries. The amount of garbage is so large that archaeology today can be expanded for the branch of 'garbageology' as William Rathje named this type of research (Zimring and Rathje 2012). But visible and instant changes to the environment are not the only ones. There are also some long run and less visible processes. Thus, there is no difference between the man standing in front of the dirty river and us. We only do not 'see' that pollution.

Yet unlike artistic images that find their way to galleries, or scientific images whose audience is academic, networked images are distributed across the global network. While the amateurization of photography and the prevalence of mobile photography have not been welcomed by professional photographers, such photography is quite important to environmental studies. In cases of a lack of images witnessing human-induced changes to the environment, amateur images can inspire and effect radical change. Thus, it behooves environmental humanities scholars to further research and support such projects of documentation and assist their further distribution to critical readers.

Literature

Barrell, John. 1980. *The Dark Side of Landscape,* Cambridge, Cambridge University Press.

Barthes, Roland. 1957 (1973). 'The Great Family of Man', in: Ronald Barthes, *Mythologies*, Paladin books, 100–102.

Batchen, Geoffrey. 1997. *Burning with Desire: the Conception of Photography.* Massachusetts, CA: MIT Press.

Batchen, Geoffrey. 2008. 'Snapshots: Art History and the Ethnographic Turn'. *Photographies* 1 (2): 121–142.

Batchen, Geoffrey. 2000. *Each Wild Idea: Writing, Photography, History.* Cambridge: MIT Press.

Barlett, Flora Mary. 2023. "Visualising the Pristine: The Role of Imagery in Local Stewardship." In *Disturbed Ecologies,* edited by Darcy White, Julia Peck, and Chris Goldie, 203. Bielefeld: Transcript Verlag.

Bear, Jordan and Kate Palmer Albers. 2017. *Before-and-After Photography: Histories and Contexts.* London: Routledge

Berger, Lynn. 2011. Snapshots, Or: Visual Culture's Clichés. Photographies 4 (2): 175–190. https://doi.org/10.1080/17540763.2011.593922.

Bermingham, Ann. 1987. *Landscape and Ideology.* Los Angeles: Berkeley University Press.

Bonneuil, Cristophe, and Jean-Baptiste Fressoz. 2016. *The Shock of the Anthropocene: The Earth, History and Us.* Translated by David Fernbach. London: Verso.

Buckley, Emma M., Brinley, Craig R., Allen, Michael Forsberg, Michael Farrell, and Andrew J. Caven. 2017. "Capturing Change: The Duality of Time-Lapse Imagery to Acquire Data and Depict Ecological Dynamics." *Ecology and Society* 22 (3): np.

Braddock, Alan C. 2014. 'Art in the Anthropocene', *American Art 28.* pp, 2–8.

Braddock, Alan. 2009. 'Ecocritical Art History', American Art 23, pp. 24–28.

Brown, Barbara. 2002. "The First Photograph." *Abbey Newsletter* 26 (3). https://cool.culturalheritage.org/byorg/abbey/an/an26/an26-3/an26-307.html.

Casey, Edward. 2001. *Earth-Mapping: Artists Reshaping Landscape.* Minneapolis and London: University of Minnesota Press.

Cheetham, Mark. 2018. *Landscape into Eco Art.* Penn State University Press.

Clark, Timothy. 2015. *Ecocriticism on the Edge: the Anthropocene as the Threshold Concept.* Bloomsbury.

Cosgrove, Denis and Stephen Daniels (eds), The iconography of Landscape: Studies in Historical Geography, Cambridge University Press, 1988.

Cosgrove, Denis. *Social Formation and Symbolic Landscape,* University of Wisconsin Press, 1988(1984)

Coughlin, Maura and Emily Gephart (eds.) 2019. *Ecocriticism and the Anthropocene in Nineteenth-Century Art and Visual Culture.* London: Routledge.

Crutzen, Paul. 2002. 'Geology of Mankind' *Nature* 414. https://doi.org/10.1038/415023a

Daniels, Stephen (eds.). *The iconography of Landscape: Studies in Historical Geography*, Cambridge University Press,

Davis, Heather and Etienne Turpin (eds.). 2015. *Art in the Anthropocene: Encounters among Aesthetics, Politics, Environments and Epistemologies.* London: Open Humanities Press.

De Duve, Thierry. 1978. 'Time Exposure and Snapshot: The Photograph as Paradox' October 5, Photography (Summer): pp. 113–125.
Demos. T. J. 2013. *The Migrant Image: The Art and Politics of Documentary During Global Crisis.* Durham/London: Duke University Press.
Demos, T. J. 2016. *Decolonizing Nature: Contemporary Art and the Politics of Ecology.* Berlin: Sternberg Press.
Demos. T. J. 2017. *Against the Anthropocene: Visual Culture and Environment Today.* Sternberg Press.
Elkins, James and Ziady DeLue, Rachel (eds.), *Landscape theory*, Routledge, 2008.
Emmelhainz, Irmgard. 2015. Conditions of Visuality U 63. Under the Anthropocene and Images of the Anthropocene to Come', *E-flux Journal.* https://www.e-flux.com/journal/63/60882/conditions-of-visuality-under-the-anthropocene-and-images-of-the-anthropocene-to-come/
Fisher, Andrew. 2013. "Photographic Scale." *In On the Verge of Photography: Imaging Beyond Representation*, edited by Daniel Rubinstein, Johnny Golding, and Andrew Fisher, 151–169. Birmingham: ARTicle Press.
Flowers et al. 'Snap Decisions. Assessing Participation and Data Quality in a Citizen Science Program using Repeat Photography', *Citizen Science: Theory and Practice.* Nov 2023.
Garrard, Rodney, and Mark Carey. 2017. "Beyond Images of Melting Ice." In *Before-and-After Photography: Histories and Contexts*, edited by Jordan Bear and Kate Palmer Albers, 101–122. Bloomsbury.
Gombrich, Ernst. 1964. *The Story of Art.* London. The Phaidon Press.
Hand, Martin (2012), UbiquitousPhotography. Cambridge: Polity Press
Hauser, *Kitti, Shadow Sites: Photography, Archaeology, and the British Landscape 1927–1955,* Oxford University Press, 2007.
Hedin, Gry and Ann-Sofie N. Gremaud. 2018. *Artistic Visions of the Anthropocene North.* Routledge.
Hell, Julia and Schönle, *Ruins of Modernity*, Duke University Press, 2010.
Jurgenson, Nathan. 2019. *The Social Photo: On Photography and Social Media.* Verso.
Kastner, Jeffrey and Brian Wallis 2010. *Land and Environmental Art.* Phaidon Press.
Kember, Sarah. 2017. "After the Anthropocene: The Photographic for Earthly Survival." *Digital Creativity* 28: 348– 353.
Klein, Christopher. 2012. 'The Great Smog of 1952'. *History.* Dec. https://www.history.com/news/the-killer-fog-that-blanketed-london-60-years-ago
Krauss, Rosalind. 1982. 'Photography's Discursive Spaces: Landscape/View', *Art Journal* 42 (4, The Crisis in the Discipline): 311–319.

Lister, Martin. 2014. 'Overlooking, Rarely Looking and Not Looking' in Jonas Larsen, Mette Sandbye (eds.) *Digital Snaps: The New Face of Photography*. I. B. Tauris.

Mirzoeff, Nicholas. 2014. 'Visualising the Anthropocene' *Public Culture* 26 (2): 213-232.

Mitchell, WJT. *Landscape and Power*, University of Chicago Press, 2002.

Niederer, Sabine. 2018. *Networked Images: Visual Methodologies for the Digital Age*. Amsterdam: Amsterdam University of Applied Sciences.

Nickel, Douglas R. 1998. Snapshots: The Photography of Everyday Life, 1888 to the Present. San Francisco: San Francisco Museum of Modern Art.

Peraica, Ana. 2021. "Photographocene: The Past, Present, and Future in the Photography of the Environment." *Philosophy of Photography* 11 (1–2): 99–111.

Robertson, Kirsty. 2016. "Plastiglomerate," e-flux 78 (December).

Rubinstein, Daniel and Sluis, Katrina. 2008. 'A life more photographic', Photographies,1:1,9 — 28. https://doi.org/10.1080/17540760701785842

Shanks 2016. *The Archaeological Imagination*, London: Routledge.

Sheppard, Stephen R. J. 2005. 'Landscape visualisation and climate change: The potential for influencing perceptions and behaviour'. *Environmental Science and Policy* 8(6): 637-654.

Spence, Patricia and Jo Holland (eds). 1991. *Family Snaps: The Meanings of Domestic Photography.* Virago Press Limited.

Stengers, Isabelle. 2015. "Accepting the Reality of Gaia: A Fundamental Shift?" In *The Anthropocene and the Global Environmental Crisis: Rethinking Modernity in a New Epoch*, edited by Clive Hamilton, Cristophe Bonneuil, and François Gemenne, 29–46. Abingdon and New York: Routledge Environmental Humanities Series

Reiss, Julie. 2018. *Art, Theory and Practice in the Anthropocene*. Vernon Press.

Webb, Robert H. 2010. *Repeat Photography, Methods, Applications in the Natural Sciences* Washington DC: Island Press.

Weintraub, Linda. 2018. *Eco Materialism and Contemporary Art*. Bristol: Intellect Books.

Wells, Liz. 2022. *Land Matters: Landscape Photography, Culture and Identities*. Abingdon and New York: Taylor and Francis.

Zalasiewicz, Jan. 2008. *The Earth after Us*. Cambridge: Oxford University Press.

Zimring, Carl. A. and William L. Rathje. 2012. *Encyclopaedia of Consumption and Waste: The Social Science of Garbage*. Thousand Oaks: SAGE.

CHAPTER 8

Losing Our Heads: Finding Bodies in the Digital Anthropocene Through Contemporary Painting Practice

Rachel Magdeburg

> When the organic bond between nature and human beings is severed, nature becomes a passive object rather than an active partner (Merchant 1989: 120).

Preface

The Anthropocene was a proposed new epoch for the geological timescale, prompted by human (not all) interventions on planetary biophysical systems, and although the scientific proposal has recently been formally rejected, the concept retains resonance and meaning. The

R. Magdeburg (✉)
Lecturer in Fine Art (Painting), Manchester Metropolitan University, Manchester, UK
e-mail: R.Magdeburg@mmu.ac.uk

D. Binns and R. Najdowski (eds.), *Confronting the Climate Crisis*, Palgrave Studies in Media and Environmental Communication,
https://doi.org/10.1007/978-3-031-89606-4_8

Anthropocene has resulted in a condition of '*rupture*' (Hamilton 2016: 94, original emphasis), propelled by western humanist dominion over ~~nature~~ (and *sous-rature* is applied herein to strike at the inadequacy of this term and its separation from culture). However, this worldview falters when faced with climate volatility, anthropogenic evidence in the geological record and humans playing host to the SARS-CoV-2 virus. The result is a hubristic loss of face and losing one's head to anxiety (Robbins and Moore 2013), grief (Head 2016), and 'solastalgia'—the fear of losing one's home (Albrecht et al. 2007).

This chapter expounds on the expression of losing one's head. I speculate that the severance of mind and body in Cartesian dualism, and the modern schism of subject and object, can metaphorically and visually be encapsulated in artistic representations of decapitation. I argue emoji (using the singular form in agreement with a recent argument, see Meyer 2016) are contemporary beheadings, which can recover bodies lost in the digital interface, convey the pandemic's troubling of proximity and touch, and express Anthropocene wounds through affect. Utilizing art practice as research and a semio-materialist approach—which considers the semiotic as material and matter as semiotic—I give body to emoji in a series of paintings to demonstrate they are not lifeless signs but lively and can rearrange and question the subject and object split.

Rupture, Dualism, and Severance

The Anthropocene's universalizing logic is 'structured to sever the relations between mind, body, and land' (Davis and Todd 2017: 761). Crutzen and Stoermer's (2000) originary Anthropocene proposal emphasized the noösphere—a layer of thought '*above* the biosphere and geosphere' (Davis and Todd 2017: 768, original emphasis)—replicating Euro-western severance of thought from the rest of biota and land. The severance of mind from matter in Cartesian dualism held preference for the mind. René Descartes distinguished between the mind, what he called 'I', and the body. The mind was not located in the body; it was a 'thinking thing', an incorporeal substance, liberated from the mechanical body, a 'non-thinking thing', and the soul existed without any material thing (Descartes 2013: 109). The mind was indivisible, whereas parts of the body could be chopped off with no impact on the mind (Descartes 2013: 119). Thus, the human was severed in two—an immaterial mind and a mechanical body. The 'Cartesian cut' brought an innate separation between subjects and objects, whereas subjects and objects in the 'agential cut' are not a priori relata but emerge through 'intra-action' (Barad 2003: 815).

Descartes' ideas persevered, establishing a western 'ontological rupture' whereby ~~nature~~ was constructed as external and primordial, and intellectualized life and work, urban (Fitzsimmons 1989: 107). The 'Enlightenment antimonies' (Fitzsimmons 1989: 108) of mind and matter extended to the modern bifurcation of subject and object (Latour 1993). Once the subject is separated from the object, it downgrades it and forgets how much it is also an object (Adorno 2005), leading to humans becoming 'the lords and possessors of nature' (Descartes 2008: 102). To intervene in ~~nature~~, it had to be conceived as an object involving 'a separated mind observing separated matter; man looking at nature' (Williams 1980: 77). This seeded the Anthropocene condition, yet paradoxically this concept brings an 'utter confusion between objects and subjects' (Latour 2014: 8). The partition of subject and object reverberates across academic disciplines and epistemologies, lying at the heart of looking and displaying (Bal 1996). It also infects the philosophy of language, by differentiating between lifeless spoken and written signs and animating thoughts and ideas (Hagberg 2005). Therefore, contemporary painting and emoji are intimated.

The Scientific Revolution's mechanistic worldview (Merchant 1989) required the removal of the subject from the scientific object resulting not in a pure object, according to Adorno, but a false one. The object was not a subject-less remainder after the subject had been subtracted. Depersonalizing knowledge for objective purposes retained nothing but 'the *caput mortuum* of objectivity' (Adorno 2005: 253). Adorno's objectivity-residue concept figured the alchemical symbol *caput mortuum*, Latin for 'dead head'. This was the worthless 'refuse scoria, recrement' after a sample. Graphically, the *caput mortuum* symbol 'Death's Head' (Elkins 1999: 141) (Fig. 8.1) bares resemblance to contemporary emoji (Fig. 8.2).

If accepting this proposition, by boiling out subjectivity to satisfy scientific rigour, ironically results in a subjective visual expression of affect; not objective sediment, but a graphic approximation of the subject *and* a vehicle for subjective sentiment, a move from dead head to charismatic emoji. Emoji are affective, co-produce agency and hold traces of mental (ill)health amplified by the Anthropocene's ecocide (Mcphie 2019). Mental health is not bound inside heads but extends environ(mentally),

Fig. 8.1 *Alchemical Symbol for Caput Mortuum*, screenshot. Unicode character code U + 1F74E. Noto Sans Symbols, Google Fonts. Copyright 2022 The Noto Project Authors (https://github.com/notofonts/symbols). This Font Software is licensed under the SIL Open Font License, Version 1.1

Fig. 8.2 *Face with Open Mouth*. Unicode character code U + IF62E. Mariella Steeb, designed by OpenMoji—the open-source emoji and icon project. License: CC BY-SA 4.0

distributed into objects such as emoji, which play a role in conveying and producing mental health, leaking into the digital environment as 'extended face[s]' (Mcphie 2019: 154). Therefore, emoji can express Anthropocene wounds and are manifestations of them (Fig. 8.3).[1]

Decapitation in Western Civilization and Early and Modern European Painting

The rupture between intellect, subject and mind, from matter, object and body, can metaphorically and visually be expressed in the split of head from body in decapitation. Decapitation is 'among the most ancient, widespread, and enduring of cultural practices' (Janes 2005: 10). It occurred in early hominid species, distinguishing ours from nonhumans

Fig. 8.3 Rachel Magdeburg, *Toppling*, 2019. Watercolour on paper. 42 × 30 cm. © Rachel Magdeburg

(Janes 2005) where it is a by-product or accident (Maxwell 1998), and continues today in acts of brutality and capital punishment (Amnesty International 2021). Regina Janes predicts that when 'a discourse is in crisis, a severed head is likely to turn up' (Janes 2005: xiii). Julia Kristeva reiterates this by adding that 'beheadings appear in the midst of schisms and discord' (Kristeva 2012: 104). The ecological, sociopolitical, and philosophical crisis of the Anthropocene means a head is imminent.

Although decapitation implies finality, in European cultural history insubordinate, mocking and moving heads motion. In late antiquity, *cephalophores* were martyrs and saints who carried their heads after decapitation to their places of rest (Baert et al. 2013). Charlotte Corday blushed after guillotine execution when her face was struck (Camus 1963; Janes 2005; Kristeva 2012), encouraging physicians to explore 'the autonomous life of a head' (Kristeva 2012: 96). The bouncing decapitated head taunts authority's stilling intention (Janes 2005) in eighteenth-century

English literary works. In contemporary artist Ed Atkins' video installation *Ribbons* (2014) at the Serpentine Gallery, London, his CGI avatar Dave's head ricochets infinitely down a flight of stairs. Yet despite evoking subjectivity, decapitated heads were also condemned as objects. Arguing for the abolition of the death penalty in Camus' essay 'Reflections on the Guillotine', he described how the criminal was treated and called a 'parcel' by officials. The parcel was 'inert matter', a 'thing', an 'animal' (Camus 1963: 153), reducing the subject (guilty or not) to 'being an object' (Camus 1963: 154), waiting to be posted to the guillotine. In guillotine executions, a denial of physicality was claimed, purportedly bringing a humane contactless operation, taking 'life almost without touching the body' (Foucault 1995: 13). Non-corporal punishment continued contactless retribution, where it was no longer the body that was penalized, but the soul's 'faceless voices'. The punitive justice system attacked this 'bodiless reality' by disciplining minds (Foucault 1995: 17). This could be argued to persist in digital technology's dematerialized pretence of conducting contactless operations, which ratchet up depression, suicide and stress (Berardi 2009, 2010; Steyerl 2012, 2015). The Internet is also without a head, it 'has no face' (Aranda et al. 2015: 5), and its 'headlessness' (Janes 2005: 187) and 'acephalous media' (Joselit 2011: 86)[2] assume an egalitarian façade (Aranda et al. 2015) as it is without a top singular authority. The Internet provides knowledge but also absorbs it, harvesting our attentions, emotions, time and vision (Aranda et al. 2015). In the West, the sovereign head may have shrunk, but headless networked media governs through signs and minds as a 'sovereignty of data' (Hu 2015: xvi).

Since the Cinquecento, decollation has been a frequent motif in European painting drawing on scripture and mythology. Decapitation is indicated in western representation itself through duplication, reflection, simulacrum, mirror and imprint (Kristeva 2012). Death was rendered visible through the 'capital cut' of the '*graphic line*' (Kristeva 2012: 84, 55, original emphasis), which incises space and forms. In a series of emoji paintings called *Why the Sad Face?* (2019–2020) I used a plate to stencil a 'graphic line', cutting a circular head through extracting paint with water, recalling the platter John the Baptist's head was deposited on (Fig. 8.4).

A sub-genre of decapitation iconography is the 'self-portrait en *décapité*' (Panofsky 1969: 43, original emphasis) where artists used their own physiognomy for the fated head. De Koomen asserts this was designed to create a 'jocular effect' by artists representing themselves

Fig. 8.4 Rachel Magdeburg, *Taking out Sunset*, 2019. Watercolour on paper. 42 × 30 cm. © Rachel Magdeburg

dead in their own work (De Koomen 2013: 218), self-deprecating as a 'pathetic object' (Janes 2005: 13) whilst mastering this feat (De Koomen 2013). Contemporary painters using self-effacement and self-deprecation by decapitating subjects (Herbert and Cooke 2010) reduce the status of *anthropos* in the Anthropocene by beheading them, whilst paradoxically authoring this shrinkage through artistic autonomy. Contemporary painters use decapitation in politicized, racialized, humorous, and art historically inflected ways, ratifying this as steadfast iconography. My artistic practice offers emoji as suitable decapitation imagery for the Anthropocene.

Unruly human heads pale in comparison to nonhuman capacity to regenerate heads and body parts (Elliott and Alvarado 2013; Gemberling et al. 2013; Sprecher et al. 2015; Vogg et al. 2019). Medusa's bloody neck spawned twin progeny Pegasus and Chrysaor and despite changes over time to Medusa's physiology and media used, what persists is her 'full face', roundness (Vernant and Vidal-Naquet 1990: 191) and frontality (Frontisi-DuCroux 2003; Glennon 2017). This frontal regard, often portrayed with tongue sticking out (Stafford 1972; Vernant and Vidal-Naquet 1990), is an affront to the 'sky-gazing Anthropos' (Haraway 2016: 10). Medusa's head is only tolerable as 'an *eikon*'; by looking at her in the reflection of a mirror (Frontisi-DuCroux 2003: 264, original emphasis). Medusa as *eikon* (Fig. 8.6) finds visual correspondence with an on-screen digital icon (Fig. 8.5).

Medusa stares at her onlooker, hastening a confusion between subject and object (Bal 1994) and placing the subject also under the gaze (Foster 2003). This slippage between 'I' and 'thou' occurs in Caravaggio's painting *Head of Medusa* (1597–1598), which is represented on a convex shield. Likely to be a self-portrait *en décapité,* this painting presupposes the use of a mirror, complicating the subject/object relation (Marin 1995; Bal 1996). Medusa is germane to the Chthulucene,

Fig. 8.5 *Face with Tongue*. Unicode character code U + 1F61B. Mariella Steeb, designed by OpenMoji—the open-source emoji and icon project. License: CC BY-SA 4.0

Fig. 8.6 Ergotimos, Gorgoneion Terracota Stand, c.570 BCE. Terracotta, 5.7 × 9 cm. The Metropolitan Museum of Art, New York. Public Domain

Donna Haraway's (2016) alternative to the Anthropocene that riffs yet differs from H.P. Lovecraft's fictional monster Cthulhu. Haraway references the mythological story of Perseus laying Medusa's decapitated head on seaweed, spilling her blood into the sea which petrified as coral (Ovid 1717). Coral bleaching is a metonym for ecological disaster and the Anthropocene, with tropical sea surface temperature rises likely to increase bleaching events (Hughes et al. 2018) (8.7).

Fig. 8.7 Rachel Magdeburg, *Watching Perseus*, 2019. Watercolour on paper. 42 × 30 cm. © Rachel Magdeburg

Decollated Emoji Find Bodies

Arguably the most ubiquitous contemporary severed heads are emoji face characters. Emoji are adopted pervasively, globally, and according to Vyvyan Evans, they underscore what it means to be human (Evans 2017). Contemporary artists have used emoji in artworks, they form part of the Museum of Modern Art's permanent collection in New York (Hess 2016) and feature in art exhibitions (*Emoji Art & Design Show* 2013; *Zoe Beloff | Emotions Go to Work* 2019).

It is widely accepted that digital text lacks non-verbal cues (Krohn 2004; Alshenqeeti 2016; Lucas 2016; Evans 2017; Riordan 2017) such as body language, paralinguistics and speech prosody used in face-to-face communication (Evans 2017). Sending an SMS, MMS, social media message or e-mail is devoid of eye contact, vocal intonation, body language and facial expression present in face-to-face communication. Emoji reinforce the intention (Dresner and Herring 2010; Evans 2017) and emotion of a text message that is otherwise lost (Evans 2017). Emoji become 'non-verbal surrogates' (Alshenqeeti 2016: 60) and alter the 'intensity of affect' of a message (Riordan 2017: 4), transplanting real-world heads and bodies with digital heads.

During the COVID-19 pandemic, there was an escalation in online communication and emoji use (Kelly 2020), yet ironically the pandemic delayed the release of new emoji in the 14.0 version of the Unicode Standard (The Unicode Consortium 2020). Physical and social distancing restrictions decreed during periods of societal lockdown deprived people of intimate touch, which is associated with higher levels of anxiety and feelings of loneliness, which are detrimental to psychological well-being (von Mohr et al. 2021). Therefore, if emoji replace bodies in digital text, they can also help communicate emotions compounded by loss of physical touch in the pandemic.

The painting *Hole of Nature* (Fig. 8.8) depicts the emoji 'Face with Medical Mask'. This emoji was one of the highest-ranked characters responding to the pandemic, alongside 'Microbe' (Broni 2020). In a Twitter (now X) data sample, two laughing emoji were used most in conjunction with the text 'Coronavirus', 'COVID-19', or 'COVID19'. It has been claimed that this was due to their previous popularity on the platform (Broni 2020). Nonetheless, emoji 'Face with Tears of Joy' and 'Rolling on the Floor Laughing' may also have expressed the pandemic through humour (Evans 2020).

Fig. 8.8 Rachel Magdeburg, *Hole of Nature*, 2019. Watercolour on paper. 61.6 × 45.2 cm. © Rachel Magdeburg

There are no emoji specific to climate change, but artists Marina Zurkow and Viniyata Pany have attempted to remedy this by designing disaster and resilience sets of Climoji (Climoji n.d.). British energy company EDF launched 'Net Zero' emoji to help customers discuss climate change (EDF 2020), which could cynically be perceived as greenwashing. Emoji are affective and informational capitalism makes a bid to 'instrumentalize, analyse, monetize, and standardize affect' (Stark and Crawford 2015: 8). Emoji are prone to 'affective indeterminacy' and interpretative pliability due to the discrepancy between their iconographic and pictographic readability which can be exploited and co-opted by capitalism (Stark and Crawford 2015: 6).

Franco 'Bifo' Berardi's notion of semiocapitalism (Berardi 2009, 2010), based on 'general intellect' and taken from 1990s Italian post-operaist thinkers, defines the interaction between cognitive labour, language, and psycho-economics. In post-Fordist neo-liberal societies, the product is info-commodity and information operates semiotically in the accelerated 'emotional sphere' (Berardi 2010: 1) of networked technologies. Subjectivity and human life become products, where it is no longer bodily labour that is corralled but 'affects and desires' (Graw 2018: 252). Affect is compromised by late capitalism (Ammer 2015) and capital itself is semioticized (Steyerl 2012). Semiocapitalism 'takes the mind, language and creativity as its primary tools for the production of value' (Berardi 2009: 21) exploiting semiotic aspects of digital space. Although standardized by Unicode, representing neo-liberal lifestyles (Stark and Crawford 2015) and until recently a homogenous racial set (Tan 2015), emoji can be subversive. Unofficial editions have arisen (Dusenbery 2015), they can circumvent social media restrictions (Highfield and Leaver 2016), use humour (Evans 2020), miscommunication (Lucas 2016) and mobilize protest (Schraer 2022). As a semiotic system brought into painting, emoji can deliver ambivalent messages to avoid semiocapitalism and stand in for things other than those they picture (Bonnington 2015).

Emoji and Contemporary Painting

Recently, there has been a (re)emphasis and turn to the figure in contemporary art and painting (Mullins 2015; Della Monica 2016; Graw et al. 2011; Deitch et al. 2019; *Radical Figures: Painting in the New Millennium* 2020; *Condition Humaine* 2021; Roelstraete 2022), although I would argue the figure never completely vanished. In the Anthropocene, humans should figure in visual art but often do not (Magdeburg 2018). The post-anthropocentric reaction to the Anthropocene backfires when excluding the human figure, only redoubling dualism. Instead of bypassing the human, I use emoji as surrogates in contemporary painting, acting in the same way as other artistic depictions of the human figure, which are 'stand-in[s] for aspects of the human condition' (Mullins 2015: 10).

Recent contemporary paintings reveal how the human figure is depicted disfigured, fragmented, cartoon-like and caricatured (Searle 2020; Roelstraete 2022). The notion of the genuine, perfect, whole, white, behaving body in painting has been abandoned, and instead, 'radical figures' are never quite complete (Whitechapel Gallery 2020b: n.pag.). Recent developments in art also endow human characteristics to nonhuman objects, something capitalism makes good use of in products that adopt human traits. Therefore, artwork doing the same could coerce this doctrine (Graw 2018). For the conference *Art and Subjecthood: The Return of the Human Figure in Semiocapitalism*, Michael Fried's 1967 essay 'Art and Objecthood' was twisted into art and subjecthood, where objects were given human personalities, becoming 'quasi-subjects' (Graw 2011: 12). Subjectivity and agency were redistributed to nonhuman objects in Rachel Harrison's installation *Trees for the Forest* (2007), where the artwork seemed to have no need for a viewer and was granted 'quasi personality' (Blom 2011: 70). Subjectivities are buried into represented objects which act as surrogates and crypto-self-portraits. Paul Gauguin famously ordered sunflower seeds from France to his Tahitian base. Once cultivated, Gauguin painted these as 'surrogate portraits' (The National Gallery 2019: n.pag.) of his late friend and rival Vincent Van Gogh; objects for subjects, living things for dead artists, flower heads for human heads. In turn, Van Gogh painted two empty chairs to stand in for himself and Gauguin, and dead sunflowers, which correlated with his mental torment and the severance of his ear, bringing a 'rupture of personal homogeneity' (Bataille 1985: 68).

Why the Sad Face? series were stirred by noticing emoji reactions to my paintings disseminated on social media. Painting loses material body online, becoming retinal and transformed into light (Kelsey 2015). Emoji reactions to shared content 'compress[...] affects into icons', turning responders into producers (Kelsey 2015: 269). Recent contemporary paintings ape digital processes (Whitechapel Gallery 2020a). I use post-digital or post-internet painting to evoke digital imaging editing aesthetics such as digital colour gradation (Bacon 2016). This effect is produced using gradated watercolour wash techniques, transferring visual displays from the 'sphere of liquidity' and 'fluid media space' (Steyerl 2015: 17) to analogue liquid watercolour. Sharing completed paintings online establishes network painting, whereby objects are carried 'from one historical, topographic, or epistemological position to another (and back again)' (Joselit 2015: 173).

Networks can simultaneously be 'like a subject and an object—a person and a picture' (Joselit 2015: 174). Consequently, as a painting becomes more subject-like, the artist becomes more object-like: '[t]he subjectobject of painting [...] responded to modernity's tragicomic condition:

the subject's transformation into an object' (Joselit 2015: 179). Writing about the website Contemporary Art Daily, Michael Sanchez conceived the network as ambivalent towards nonhuman and human distinctions, between subjects and objects. Things become lively when interacted with at speed, circulating further responses (Sanchez 2011). Sanchez appears to be saying the network itself is a quasi-subject, making objects of its users. The uploading of paintings onto a digital network releases the artist's agency to the circuitry of imagery, but the risk is being turned into an object: 'the theoretical and practical tendency to subjectivize things and art objects is complemented by a continuing trend: artists are turning into virtual objects' (Graw 2018: 155). This sits in relation with Michel Serres' assertion that a quasi-object is not an object at all but an activator of subjects (Serres 1982) (Figs. 8.9, 8.10).

Some of my emoji paintings wield expressive brushstrokes (Fig. 8.11) as insignia to the illusion of the modern autonomous subject and are 'calculated spontaneity' (Hochdörfer 2015: 16). Expressive mark-making and gesture are a cipher for the pursuit of self-reassurance during times of technological revolution (Hochdörfer 2015), but expressionism struggles to be genuine after postmodern painting revealed the limits of subjectivity (Kröner 2015). Now the expressive mark is not an expression of the artist's subjectivity but an illusion of authenticity and immediacy, a rhetorical mark, a material signifier customized '*as an effect*' (Graw 2018: 52, original emphasis), engineered and premediated. Joselit argues that painting's task is to reorder the dialectic between subjects and objects through the painterly gesture, what he calls 'subjectobject'. The Impressionist's painterly mark was subjectobject as it was concurrently 'an objective optical sensation and a subjective temperament' (Joselit 2015: 169). Therefore, painterly expressive marks as subjectobject are no longer genuine but can reshuffle subjects and objects.

Why the Sad Face? paintings pursue cliché as recourse to the crisis of the subject, which has been related to past crises in painting (Hochdörfer 2015). Formalism attempted to purge mass culture, technology and cliché from painting (see Greenberg [1965] 1982), disregarding 'the option of a body *picture* and pictorial body [...] *inside* a society dominated by clichés' (Ammer 2015: 87, original emphasis). In 1966, Lucy Lippard mounted the exhibition *Eccentric Abstraction* in New York to counter the modernist, reductive account of art. Using painting as its conceptual template, the exhibition excluded actual painting. Curator Manuela Ammer's corrective to Lippard's omission of painting was

Fig. 8.9 Rachel Magdeburg, *You have a bit of chocolate...*, 2019. Watercolour and acrylic collaged on paper. 42 × 30 cm. © Rachel Magdeburg

the concept 'eccentric figuration', exampling figurative painters of the same period and celebrating the materiality and corporeality of painting as a 'picture-body praxis' (Ammer 2015: 87). Eccentric figuration is where bodies in painting mobilize cliché, fantasy, technology, caricature, cartoonish exaggeration, affect and narrative (Ammer 2015). Taking Nicole Eisenman's affective painting *Bloody Orifices* (2005), Ammer understands the depicted (decapitated) head as 'leaking colour' and 'dispensing paint' (Ammer 2015: 85) like a body, dissolving back into the material that made it. The picture and pictorial body concur in my emoji paintings as they weep and drip paint into and out from the impregnated body of the paper (Fig. 8.12). Similarly, artist Maria Lassnig's 'body-awareness' paintings of the 1960s oscillate between subject and object

Fig. 8.10 Rachel Magdeburg, *Palm Shaking*, 2019. Watercolour on paper. 42 × 30 cm. © Rachel Magdeburg

(Ammer 2015). The painting substrate becomes body-like as Lassnig's self-portraits emerge from it and the paintings were made whilst the artist was lying on the support. The 'affective qualities of cliché' in Lassnig's paintings (Ammer 2015: 94) are the merits of emoji also, and instead of cliché being omitted, it should be visualized '*as compromised*' (Ammer 2015: 95, original emphasis). In my paintings compromised emoji are pictured with malfunctioning umbrellas (Fig. 8.13), obscured behind trees (Fig. 8.14), and sliding off the page (Fig. 8.15).

The ultimate clichéd and compromised subject losing their head, is the figure in Edvard Munch's painting *The Scream* (1893), vectorized also as an emoji. Munch's overly commoditized and appropriated image begs the question, 'how could anyone continue painting once all subjects, all

Fig. 8.11 Rachel Magdeburg, *Jupiter and Io*, 2019. Watercolour on paper. 42 × 30 cm. © Rachel Magdeburg

means of expression, were infected with cliché: once all things authentic were contaminated by their surrogates?' (Hochdörfer 2015: 21). My version of the screaming subject (Fig. 8.16) is a two-colour gradation that cuts its anguished head out from the painting's ground. Emoji are clichéd but not strictly 'poor images' (Steyerl 2012). As vectors, their resolution does not deteriorate as easily as pixels. Their poverty lies in being transferred into painting, as manual brushstrokes and wash lose the crispness of digital imagery. Sharpness and exactness are replaced by wobbly lines, clumsy touches and un-flush gradations. Imprecision amounts to a 'castration of the author' (Steyerl 2012: 35); therefore, I have lost my head.

Fig. 8.12 Rachel Magdeburg, *Dormouse Eyes*, 2019. Watercolour on paper. 61.6 × 45.2 cm. © Rachel Magdeburg

How can emoji as decapitated heads be brought to life? Isabelle Graw has argued that paint gives 'human proxies' their aliveness (Graw 2018: 241). She suggests painting possesses liveliness through 'vitalistic fantasies' (2018: 20) that project human attributes onto seemingly lifeless material and acquire the qualities the living have—animation, liveliness, subjectivity. Painting's affective charge for Graw is achieved through its semiotics understood materially: corporeal pigments connect to the living earth; paintings are conjoined to the lively artist through indexical brushstrokes and material traces of artistic labour, and appropriated material from the 'real' world is inserted into paintings. New and vibrant materialist, speculative and agential realist philosophies and actor-network theories highlight the agency of objects. However, this is not a one-way process. The artist becomes more alive through interaction during

Fig. 8.13 Rachel Magdeburg, *Driving Rain*, 2019. Watercolour and acrylic on paper. 42 × 30 cm. © Rachel Magdeburg

making, allowing another 'voice' to appear when artistic intention is not predetermined, and the artist is ventriloquized by the artwork (Goldblatt 2005). Despite sounding like a Pygmalion wet dream, if painting is considered to have liveliness, it has also been described as having a 'corpus' (Hochdörfer 2015: 21), being 'body-like' (Payen 2004: 3), and painters Deb Covell, Lynda Benglis, Alexis Harding and Pavel Büchler have explored the material skin of paintings.

Post-mortem

The rupture of mind and matter, ~~nature~~ and culture, and subject and object, can be analogous to the separation of heads and bodies in visual representation. Decapitation is a widespread motif in early modern,

Fig. 8.14 Rachel Magdeburg, *In the Way Tree*, 2019. Watercolour on paper. 42 × 30 cm. © Rachel Magdeburg

modern and contemporary western painting. In my practice, I propose emoji as apt icons for decapitation iconography. Emoji act as proxies for the human, resisting the eradication of the subject in the Anthropocene without retorting to the human figure per se. Emoji replace bodies lost to digital text, thus are not complete decapitations. They do this by possessing affective capacities that stand in for body language, vocal stress, facial expressions and gesture. However, their affective capacity is imperilled by semiocapitalism's siphoning of affect, prompting painting to seek additional strategies: cliché as compromised; eccentric figuration; painterly marks as subjectobjects; and finding body in materiality and the indexical trace of the artist. Emoji characters are nonhuman surrogates and can express the wounds of the Anthropocene through humour and

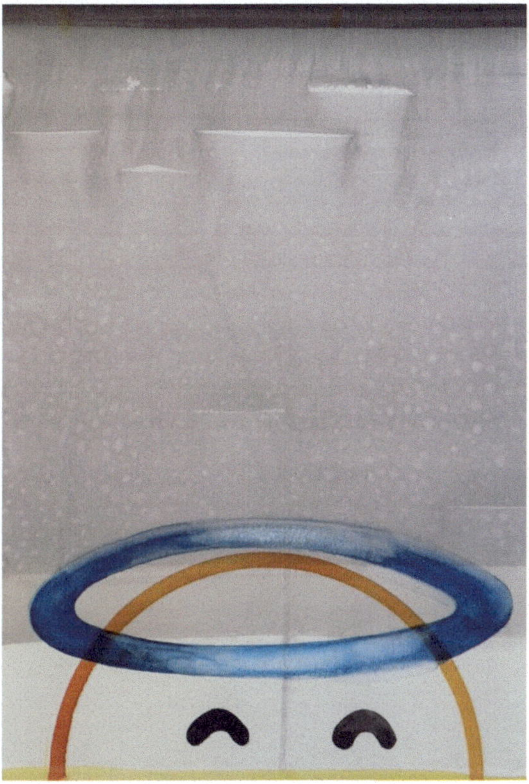

Fig. 8.15 Rachel Magdeburg, *Dropping Halo*, 2019. Watercolour on paper. 42 × 30 cm. © Rachel Magdeburg

self-deprecation. In painting, they dissolve into corporeal pigment *and* leech fluids. Watercolour emoji are affected by their own poor image creation: the forced spontaneity, the drip as cipher, the wonky graduated wash of analogue painting.

The relationship between heads and bodies, life and death, and subjects and objects is not clean cut; objects become subjects, subjects become objects. Objects, even paintings, acquire liveliness, gain subjecthood as 'quasi subjects', procure a body and affect the viewer. The COVID-19 pandemic has pushed many people to increased time online. The physical, psychological, and existential threats of the Anthropocene can be communicated through emoji although it remains unclear whether the

Fig. 8.16 Rachel Magdeburg, *Loom*, 2019. Watercolour on paper. 42 × 30 cm. © Rachel Magdeburg

severance of subject and object can be clotted by the Anthropocene thesis or remains haemophilic.

Notes

1. Mental health charity Mind released an Emoodji app for university students to monitor their moods and well-being.
2. Joselit is discussing a series called *Hostage Video Still with Time Stamp Price* (2005-) by artist Seth Price who screen-printed a reproduction of the decapitated head of businessman Nicholas Berg.

References

Adorno, Theodor W. (2005), *Critical Models: Interventions and Catchwords* (trans. H.W. Pickford), New York: Columbia University Press.

Albrecht, Glenn, Sartore, Gina-Maree, Connor, Linda, Higginbotham, Nick, Freeman, Sonia, Kelly, Brian, Stain, Helen, Tonna, Anne and Pollard, Georgia (2007), 'Solastalgia: The Distress Caused by Environmental Change', *Australasian Psychiatry*, 15:1, pp. S95–S98, https://doi.org/10.1080/10398560701701288. Accessed 29 April 2020.

Alshenqeeti, Hamza (2016), 'Are Emojis Creating a New or Old Visual Language for New Generations? A Socio-semiotic Study', *Advances in Language and Literary Studies*, 7:6, pp. 56–69, https://doi.org/10.7575/aiac.alls.v.7n.6p.56. Accessed 23 April 2020.

Ammer, Manuela (2015), '"How's My Painting?" (Judge Me, Please, Don't Judge Me)', in M. Ammer, A. Hochdörfer, and D. Joselit (eds), *Painting 2.0: Expression in the Information Age*, Munich, London and New York: DelMonico · Prestel, pp. 85–101.

Amnesty International (2021), *Amnesty International Global Report Death Sentences and Executions 2020,* London: Amnesty International, https://www.amnesty.org/en/wp-content/uploads/2021/05/ACT5037602021ENGLISH.pdf. Accessed 20 October 2021.

Aranda, Julieta, Vidokle, Anton and Kuan Wood, Brian (eds) (2015), *The Internet Does Not Exist.* Berlin: Sternberg Press.

Bacon, Alex (2016), 'Surface, Image, Reception: Painting in a Digital Age', Rhizome, 24 May, https://rhizome.org/editorial/2016/may/24/surface-image-reception-painting-in-a-digital-age/. Accessed 8 March 2020.

Baert, Barbara, Traninger, Anita and Santing, Catrien (eds) (2013), *Disembodied Heads in Medieval and Early Modern Culture*, Leiden and Boston: Brill.

Bal, Mieke (1994), 'Head Hunting: "Judith" on the cutting edge of knowledge', *Journal for the Study of the Old Testament*, 63, pp. 3–34, https://journals.sagepub.com/doi/abs/https://doi.org/10.1177/030908929401906301. Accessed 1 May 2020.

Bal, Mieke (1996), *Double Exposures: The Subject of Cultural Analysis*, New York and Abingdon, Oxon: Routledge.

Barad, Karen (2003), 'Posthumanist Performativity: Toward an Understanding of How Matter Comes to Matter', *Signs,* 28:3, pp.801-831.

Bataille, Georges (1985), *Visions of Excess: Selected Writings, 1927–1939* (ed. and trans. A. Stoekl), Minneapolis, MN: University of Minnesota Press.

Berardi, Franco 'Bifo' (2009), *The Soul at Work: From Alienation to Autonomy*, Los Angeles, CA: Semiotext(e).

Berardi, Franco 'Bifo' (2010), 'Cognitarian Subjectivation', *e-flux*, 20, pp. 1–8, https://www.e-flux.com/journal/20/67633/cognitarian-subjectivation/. Accessed 2 May 2020.

Blom, Ina (2011), 'Media Animism: Rachel Harrison's Living Images', in I. Graw, D. Birnbaum, N. Hirsch and Institut für Kunstkritik (eds), *Art and Subjecthood: The Return of the Human Figure in Semiocapitalism*, Berlin: Sternberg Press, pp. 65–75.

Bonnington, Christina (2015), 'At Long Last, The Taco Emoji Has Arrived!', Refinery 29, 21 October, https://www.refinery29.com/en-us/2015/06/89339/taco-emoji-is-here. Accessed 20 April 2020.

Broni, Keith (2020), 'Spread of the Coronavirus Emoji', Emojipedia, 11 March, https://blog.emojipedia.org/spread-of-the-coronavirus-emoji/. Accessed 30 April 2020.

Camus, Albert (1963), *Resistance, Rebellion, and Death* (trans. J. O'Brien), New York: The Modern Library.

Climoji (n.d.), 'About', Climoji, https://www.climoji.org. Accessed 20 September 2021.

Condition Humaine (2021), Grimm, New York, 29 October – 4 December.

Crutzen, Paul J. and Stoermer, Eugene F. (2000), 'The "Anthropocene"', *Global Change Newsletter*, 41, pp. 17–18, http://www.igbp.net/download/18.316f18321323470177580001401/1376383088452/NL41.pdf. Accessed 11 January 2018.

Davis, Heather and Todd, Zoe (2017), 'On the Importance of a Date, or Decolonizing the Anthropocene', *ACME: An International Journal for Critical* Geographies, 16:4, pp. 761–780, https://www.acme-journal.org/index.php/acme/article/view/1539. Accessed 2 November 2018.

Deitch, Jeffrey, Gingeras, Alison, Dean, Aria and Fateman, Johanna (2019), *Unrealism: New Figurative Painting*, New York: Rizzoli.

Della Monica, Lauren P. (2016), *Bodies of Work - Contemporary Figurative Painting*, Atglen, PA: Schiffer.

De Koomen, Arjan R. (2013), 'The Self-Portrait 'En Décapité': Interpreting Artistic Self-Insertion', in B. Baert, A. Traninger and C. Santing (eds), *Disembodied Heads in Medieval and Early Modern Culture*, Leiden and Boston: Brill, pp. 191–221.

Descartes, René (2008), *Discourse on the Method of Rightly Conducting One's Reason and of Seeking Truth* (trans. J. Veitch), Salt Lake City, UT: Project Gutenberg, https://www.gutenberg.org/ebooks/59. Accessed 3 March 2020.

Descartes, René ([1641] 2013), *René Descartes: Meditations on First Philosophy: With Selections from the Objections and Replies. A Latin-English Edition* (trans. J. Cottingham), New York: Cambridge University Press.

Dresner, Eli and Herring, Susan C. (2010), 'Functions of the Nonverbal in CMC: Emoticons and Illocutionary Force', *Communication* Theory, 20:3, pp. 249–268, https://doi.org/10.1111/j.1468-2885.2010.01362.x. Accessed 7 May 2020.

Dusenbery, Maya (2015), 'These New Vagina Emojis are the Best', Feministing,15 October, http://feministing.com/2015/10/15/these-new-vagina-emojis-are-the-best/. Accessed 6 May 2020.

EDF (2020), 'New emojis to help public discuss Net Zero', EDF, 7 October, https://www.edfenergy.com/media-centre/news-releases/new-emojis-help-public-discuss-net-zero. Accessed 20 September 2021.

Elkins, James (1999), *What Painting is: How to Think about Oil Painting, Using the Language of Alchemy*, New York and London: Routledge.

Elliott, Sarah A. and Sánchez Alvarado, Alejandro (2013), 'The history and enduring contributions of planarians to the study of animal regeneration', *WIREs Developmental Biology*, 2:3, pp. 301-326, https://doi.org/10.1002/wdev.82. Accessed 1 May 2020.

Evans, Vyvyan (2017), *The Emoji Code: How Smiley Faces, Love Hearts and Thumbs Up are Changing the Way We* Communicate, London: Michael O'Mara Books.

Evans, Vyvyan (2020) 'Coronavirus Emojis: Conveying Compassion and Humour With a Facemask', 4 May, https://thewire.in/culture/coronavirus-emojis-conveying-compassion-and-humour-with-a-facemask. Accessed 10 May 2020.

Emoji Art & Design Show, Eyebeam, New York, 13–14 December 2013.

Fitzsimmons, Margaret (1989), 'The Matter of Nature', *Antipode*, 21:2, pp. 106–120, https://doi.org/10.1111/j.1467-8330.1989.tb00183.x. Accessed 3 May 2020.

Foster, Hal (2003), 'Medusa and the Real', *RES: Anthropology and Aesthetics*, 44, pp. 181–190, http://www.jstor.org/stable/20167613. Accessed 14 September 2020.

Foucault, Michel (1995), *Discipline and Punish: The Birth of the Prison* (trans. A. Sheridan), 2nd ed., New York: Vintage Books.

Frontisi-Ducroux, Françoise ([1993] 2003), 'Medusa as Maker of Images', in M. Garber and N. J. Vickers (eds), *The Medusa Reader*, New York and Abingdon, Oxon: Routledge, pp. 262–266.

Gemberling, Matthew, Bailey, Travis J., Hyde, David R. and Poss, Kenneth D. (2013), 'The zebrafish as a model for complex tissue regeneration' *Trends in genetics:* TIG, 29:11, pp. 611–620, https://pubmed.ncbi.nlm.nih.gov/23927865 https://www.ncbi.nlm.nih.gov/pmc/articles/PMC3812420/. Accessed 2 May 2020.

Glennon, Madeleine (2017), 'Medusa in Ancient Greek Art, The Metropolitan Museum of Art, March, https://www.metmuseum.org/toah/hd/medu/hd_medu.htm. Accessed 23 April 2020.

Goldblatt, David (2005), *Art and Ventriloquism*, London and New York: Routledge.

Graw, Isabelle (2011), 'Introduction: When Objecthood Turn into Subjecthood', in I. Graw, D. Birnbaum, N. Hirsch and Institut für Kunstkritik (eds), *Art and Subjecthood: The Return of the Human Figure in Semiocapitalism*, Berlin: Sternberg Press, pp. 11–18.

Graw, Isabelle (2018), *The Love of Painting*, Berlin: Sternberg Press.

Graw, Isabelle, Birnbaum, Daniel, Hirsch, Nikolaus and Institut für Kunstkritik (eds) (2011), *Art and Subjecthood: The Return of the Human Figure in Semiocapitalism*, Berlin: Sternberg Press.

Greenberg, Clement ([1965] 1982), 'Modernist Painting', in F. Frascina and C. Harrison (eds), *Modern Art and Modernism: A Critical Anthology*, London: Paul Chapman in association with the Open University, pp. 5–10.

Hagberg, Garry L. (2005), 'Critical Commentary: Meaning Beside Itself', in D. Goldblatt, (ed.), London and New York: Routledge, pp. 1–32.

Hamilton, Clive (2016), 'The Anthropocene as Rupture', *The Anthropocene* Review, 3:2, pp. 93–106, https://doi.org/10.1177/2053019616634741. Accessed 20 April 2019.

Haraway, Donna (2016), 'Tentacular Thinking: Anthropocene, Capitalocene, Chthulucene', *e*-flux, 75, pp. 1–17, https://www.e-flux.com/journal/75/67125/tentacular-thinking-anthropocene-capitalocene-chthulucene/. Accessed 3 June 2018.

Head, Lesley (2016), *Hope and Grief in the Anthropocene: Re-conceptualising human–nature relations*, Abingdon, Oxon and New York: Routledge.

Herbert, Martin and Cooke, Nigel (2010), 'Nigel Cooke and Martin Herbert in Conversation', in Stuart Shave/Modern Art (ed.), *Nigel Cooke*, London: Koenig Books, pp.9-14.

Hess, Amanda (2016), 'Look Who's Smiley Now: MoMA Acquires Original Emoji', *The New York Times*, October 26, https://www.nytimes.com/2016/10/27/arts/design/look-whos-smiley-now-moma-acquires-original-emoji.html. Accessed 9 January 2019.

Highfield, Tim and Leaver, Tama (2016), 'Instagrammatics and digital methods: studying visual social media, from selfies and GIFs to memes and emoji', *Communication Research and* Practice, 2:1, pp. 47–62, https://doi.org/10.1080/22041451.2016.1155332. Accessed 20 April 2020.

Hochdörfer, Achim (2015), 'How the World Came in?', in M. Ammer, A. Hochdörfer and D. Joselit (eds), *Painting 2.0 Expression in the Information Age*, Munich, London and New York: DelMonico · Prestel, pp. 15–27.

Hu, Tung-Hui (2015), *A Prehistory of the Cloud*, Cambridge, MA; London: The MIT Press.

Hughes, Terry P., Anderson, Kristen D., Connolly, Sean R., Heron, Scott F., Kerry, James T., Lough, Janice M., Baird, Andrew H., Baum, Julia K., Berumen, Michael L., Bridge, Tom C., Claar, Danielle C., Eakin, C. Mark, Gilmour, James P., Graham, Nicholas A.J., Harrison, Hugo,

Hobbs, Jean-Paul A., Hoey, Andrew S., Hoogenboom, Mia, Lowe, Ryan J., Mcculloch, Malcolm T., Pandolfi, John M., Pratchett, Morgan, Schoepf, Verena, Torda, Gergely. and Wilson, Shaun K. (2018), 'Spatial and temporal patterns of mass bleaching of corals in the Anthropocene', *Science*, 359:6371, pp. 80–83, https://www.science.org/doi/https://doi.org/10.1126/science.aan8048. Accessed 9 October 2020.

Janes, Regina (2005), *Losing Our Heads: Beheadings in Literature and Culture*, New York and London: New York University Press, www.jstor.org/stable/j.ctt9qfx4d. Accessed 1 April 2020.

Joselit, David (2011), 'What to Do with Pictures?', *October*, 138, pp. 81–94.

Joselit, David (2015), 'Reassembling Painting', in M. Ammer, A. Hochdörfer and D. Joselit (eds), Painting 2.0: Expression in the Information Age, Munich, London and New York: DelMonico. Prestel, pp.169-181.

Kelly, Samantha Murphy (2020), 'People can't stop using these emojis during the pandemic 😷', CNN,11 May, https://edition.cnn.com/2020/05/11/tech/emoji-covid-19/index.html. Accessed 25 January 2021.

Kelsey, John (2015), 'The Sext Life of Painting', in M. Ammer, A. Hochdörfer, and D. Joselit (eds), *Painting 2.0: Expression in the Information Age*, Munich, London and New York: DelMonico. Prestel, pp. 268–270.

Kristeva, Julia (2012), *The Severed Head: Capital Visions* (Translated by Jody Gladding) New York: Columbia University Press.

Krohn, Franklin B. (2004), 'A Generational Approach to Using Emoticons as Nonverbal Communication', *Journal of Technical Writing and* Communication, 34:4, pp. 321–328, https://doi.org/10.2190/9EQH-DE81-CWG1-QLL9. Accessed 13 April 2020.

Kröner, Tonio (2015), 'Ridiculous Creatures', in M. Ammer, A. Hochdörfer and D. Joselit (eds), *Painting 2.0: Expression in the Information Age*, Munich, London and New York: DelMonico. Prestel, pp. 250–253.

Latour, Bruno (1993), *We Have Never Been Modern* (trans. C. Porter), Cambridge, MA: Harvard University Press.

Latour, Bruno (2014), 'Agency at the time of the Anthropocene', *New Literary* History, 45, pp. 1–18, http://www.bruno-latour.fr/sites/default/files/128-FELSKI-HOLBERG-NLH-FINAL.pdf. Accessed 3 March 2018.

Lucas, Gavin (2016), *The Story of Emoji*, Munich: Prestel.

Magdeburg, Rachel (2018), 'Self-Capture in the Anthropocene: The Expedition Paintings of Alexis Rockman', *View. Theories and Practices of Visual* Culture, 22, https://www.pismowidok.org/en/archive/2018/22-how-to-see-the-antropocene/self-capture-in-the-anthropocene. Accessed 1 January 2019.

Marin, Louis (1995), *To Destroy Painting* (trans. M. Hjort), Chicago and London: The University of Chicago Press.

Maxwell, Michael R. (1998), 'Lifetime mating opportunities and male mating behaviour in sexually cannibalistic praying mantids', *Animal* Behaviour, 55:4, pp. 1011–1028, https://www.sciencedirect.com/science/article/pii/S0003347297906710. Accessed 4 May 2020.

Mcphie, Jamie (2019), *Mental Health and Wellbeing in the Anthropocene: A Posthuman Inquiry*, Singapore: Springer Singapore, Palgrave Macmillan.

Merchant, Carolyn (1989), *The Death of Nature: Women, Ecology, and the Scientific* Revolution, New York: Harper & Row.

Meyer, Robinson (2016), 'What's the Plural of Emoji?', The Atlantic, 6 January, https://www.theatlantic.com/technology/archive/2016/01/whats-the-plural-of-emoji-emojis/422763/. Accessed 17 July 2021.

Mullins, Charlotte (2015), *Picturing People: The New State of the Art*, London: Thames & Hudson.

Nochlin, Linda (2001), *The Body in Pieces: The Fragment as a Metaphor of* Modernity, London: Thames & Hudson.

Ovid (1717), *Ovid's metamorphoses in fifteen books. Translated by the most Eminent Hands. Adorn'd with Sculptures* (trans. J. Dryden, J. Addison, S. Croxall, N. Tate, W. Congreve and S. Garth), London: Jacob Tonson. Available at: <https://www-proquest-com.ezproxy.wlv.ac.uk/books/ovids-metamorphoses/docview/2147693297/se-2?accountid=14685>.

Panofsky, Erwin (1969), *Problems in Titian, Mostly Iconographic*. London: Phaidon.

Payen, Sally Claire. (2004), 'Embodying practices: Aspects of time, body and place in the making of paintings', Ph.D. doctoral thesis, Brighton: University of Brighton.

Radical Figures: Painting in the New Millennium (2020) Whitechapel Gallery, London, 6 February – 30 August.

Riordan, Monica A. (2017), 'Emojis as Tools for Emotion Work: Communicating Affect in Text Messages', *Journal of Language and Social* Psychology, 36:5, pp. 549–567, https://doi.org/10.1177/0261927X17704238. Accessed 13 April 2020.

Robbins, Paul and Moore, Sarah A. (2013), 'Ecological anxiety disorder: diagnosing the politics of the Anthropocene' *Cultural Geographies*, 20:1, pp. 3–19, https://doi.org/10.1177/1474474012469887. Accessed 6 March 2020.

Roelstraete, Dieter (2022), 'This Figuring: Some Remarks on Today's Painting Crazes' *Mousse*, 11 January, https://www.moussemagazine.it/magazine/this-figuring-dieter-roelstraete-2022/. Accessed 23 January 2022.

Sanchez, Michael (2011), 'Contemporary Art Daily', in I. Graw, D. Birnbaum, N. Hirsch and Institut für Kunstkritik (eds), *Art and Subjecthood: The Return of the Human Figure in Semiocapitalism*, Berlin: Sternberg Press, pp. 53–61.

Schraer, Rachel (2022), 'The Russians using emojis to evade censors', *BBC News*, 10 March, https://www.bbc.co.uk/news/60649725. Accessed 11 March 2022.

Searle, Adrian (2020), 'Radical Figures review – sex, snakes, sprouting heads and flying burgers', *The Guardian*, 5 February, https://www.theguardian.com/artanddesign/2020/feb/05/radical-figures-review-whitechapel-gallery-london-painting-in-the-new-millennium. Accessed 4 May 2020.

Serres, Michel (1982), *The Parasite* (trans. L. R. Schehr), Baltimore, MD: The Johns Hopkins University Press.

Sprecher, Simon G., Bernardo-Garcia, F. Javier, van Giesen, Lena, Hartenstein, Volker, Reichert, Heinrich, Neves, Ricardo, Bailly, Xavier, Martinez, Pedro and Brauchle, Michael (2015), 'Functional brain regeneration in the acoel worm *Symsagittifera roscoffensis*', *Biology* Open, 4:12, pp. 1688–1695, http://bio.biologists.org/content/4/12/1688.abstract. Accessed 1 May 2020.

Stafford, Barbara (1972), '"Medusa" or the Physiognomy of the Earth: Humbert de Superville's Cosmological Aesthetics', *Journal of the Warburg and Courtauld* Institutes, 35, pp. 308–338, http://www.jstor.org/stable/750935. Accessed 18 September 2020.

Stark, Luke and Crawford, Kate (2015), 'The Conservatism of Emoji: Work, Affect, and Communication', *Social Media + Society*, 1:2, pp. 1–11, https://doi.org/10.1177/2056305115604853. Accessed 23 April 2020.

Steyerl, Hito (2012), 'In Defense of the Poor Image', in J. Aranda, A. Vidokle and B. Kuan Wood (eds), *e-flux journal Hito Steyerl: The Wretched of the Screen*, Berlin: Sternberg Press, pp. 32–45.

Steyerl, Hito (2015), 'Too Much World: Is the Internet Dead?', in J. Aranda, B. Kuan Wood and A. Vidokle (eds), *e-flux journal: The Internet Does Not Exist*. Berlin: Sternberg Press, pp. 11–26.

Tan, Monica (2015), 'Apple adds racially diverse emoji, and they come in five skin shades', *The Guardian*, 24 February, https://www.theguardian.com/technology/2015/feb/24/apple-adds-racially-diverse-emoji-and-they-come-in-five-skin-shades. Accessed 20 April 2020.

The National Gallery (2019), *The Credit Suisse Exhibition: Gauguin Portraits*, press release, London: The National Gallery, https://www.nationalgallery.org.uk/about-us/press-and-media/press-releases/the-credit-suisse-exhibition-gauguin-portraits. Accessed 23 December 2019.

The Unicode Consortium (2020), 'Unicode 14.0 Delayed for 6 Months', Unicode, 8 April, https://home.unicode.org/unicode-14-0-delayed-for-6-months/. Accessed 21 December 2021.

Vernant, Jean-Pierre and Vidal-Naquet, Pierre (1990), *Myth and Tragedy in Ancient Greece* (trans. J. Lloyd), New York: Zone Books.

Vogg, Matthias C., Galliot, Brigette and Tsiairis, Charisios D. (2019), 'Model systems for regeneration: *Hydra*', *Development*, 146:21, pp. 1–10, http://dev.biologists.org/content/146/21/dev177212.abstract. Accessed 2 May 2020.

von Mohr, Mariana, Kirsch, Louise P. and Fotopoulou, Aikaterini (2021), 'Social touch deprivation during COVID-19: effects on psychological wellbeing and craving interpersonal touch', *Royal Society Open* Science, 8:9, pp. 1–17, https://doi.org/10.1098/rsos.210287. Accessed 23 October 2021.

Whitechapel Gallery (2020a), 'Figurative Painting Now: Artists in Conversation', artist talks at Whitechapel Gallery, London, 6 February.

Whitechapel Gallery (2020b), *Major exhibition surveys new directions in figurative painting*, press release, London: Whitechapel Gallery, https://www.whitechapelgallery.org/wp-content/uploads/2019/11/RadicalFigures_pressrelease_Final-2.pdf. Accessed 4 May 2020.

Williams, Raymond (1980), *Problems in Materialism and Culture: Selected Essays*, London: Verso.

Zoe Beloff | Emotions Go to Work (2019) Firstsite, Colchester, 19 January – 22 April.

CHAPTER 9

Invisible Threats and Materialist Visibilities: *Degradations* by James Schneider and *Quiet Zone* by Karl Lemieux and David Bryant

Charlie Hewison

Gigantic waves, terrifying in their size and unstoppable force, engulf and destroy everything in their path: seemingly miniscule human figures, vehicles and boats, tall buildings, even New York's Statue of Liberty. Seen from above, with a detached, omniscient point of view, the eye of a devastatingly enormous storm focuses the spectator's attention, hypnotising them with the contrast between the apparent stasis of this circular point on screen and the absolute havoc with which it is associated. Cars float by collapsed houses on flooded streets, overtaking half-submerged or slowly bobbing household objects—lamps, televisions, bedframes, family photographs—that normally have no business being outside, let alone in water. Shaky footage, taken from a smartphone or a hand-held camera, accompanies the desperate fleeing from encroaching disaster—tsunami,

C. Hewison (✉)
CERILAC, Université Paris Cité, Paris, France
e-mail: charlie.hewison@gmail.com

D. Binns and R. Najdowski (eds.), *Confronting the Climate Crisis*, Palgrave Studies in Media and Environmental Communication,
https://doi.org/10.1007/978-3-031-89606-4_9

lava, hurricane...—or documents a ravaged landscape and the dazed, despairing faces of those who wander around trying to work out what has just befallen them...

These types of images are very often used in mainstream fiction film to depict climate catastrophe. Big budget apocalyptic 'cli-fi (climate fiction)'[1] films, such as Roland Emmerich's *The Day After Tomorrow* (2004), Brad Peyton's *San Andreas* (2015) or *Geostorm* by Dean Devlin (2017), all employ aesthetic and narrative tropes that depict climate catastrophe as *spectacular, hypervisible* events. It is not only in blockbuster Hollywood disaster films, however, that we find similar recourse to the hypervisible or the spectacle to communicate the effects of global warming, but also in documentaries, journalistic reports and eco-activist videos and *cinetracts*. One need only watch the various videos uploaded to YouTube by *Extinction Rebellion* to find similar aesthetic strategies that situate the issue of global warming in terms of spectacle and the 'ecological sublime' (Hitt, 1999). Alongside videos of activists and experts from around the world talking directly to the camera, and others which document specific protests, those that are explicitly produced to serve as political pamphlets often include sweeping panoramas of desolate landscapes, time-lapse footage showing the effects of deforestation from a birds-eye view, slow motion footage of avalanches, all set to inspiring oratory and stirring dramatic music.

Both in fiction film and eco-politically motivated documentary or militant video pamphlets, thus, it seems that in order to engage the audience with the potential and actual effects of climate change, hypervisible spectacle is privileged as the main aesthetic strategy. Spectacular events are brought to the foreground, matched and supplemented by grandiose cinematic effects that work to further elicit emotional response in the spectator: time-lapse, drone or aerial footage, shaky handheld cameras, hazy and squint-inducing brightness to make us feel scorching heat or blue-tinted sharpness to evoke icy cold, etc.

In the wake of the events of the last decade—the bushfires in Australia, Greece, Siberia and California, storms in Haiti, Hurricanes Harvey and Katrina in the United States, the floods in Southeast Asia and various other countries around the world, the numerous extreme heat waves including the 'Angry Summer' of 2013 in Australia, etc.—it is true that the spectacular representations of climate change that we can see in these films and videos are not necessarily misrepresentations as such: the ecological disaster that we are facing often can and does express itself in spectacularly catastrophic ways. However, audio-visual representations of these catastrophes have become ‘hypervisible’ per Jean Baudrillard in his writings on postmodern society, pornography and media (Baudrillard, 1990: 28–36; Baudrillard 1996) and that has been taken up more recently in postcolonial, feminist and queer theory (Hunnicutt 2020; Fleetwood 2011; Pidduck 2011). For this constellation of theories, ‘hypervisibility’ refers to an excessive visibility given to specific subjects, events or bodies, an overrepresentation which often has the effect of obscuring both the specific causes of that which is represented (severing the representation from the real to which it apparently refers), and those subjects which are not representable in the same way. The overrepresentation of ecological catastrophe as spectacular event seems to have similar effects, narrowing the scope of how we engage with the challenges of ecological devastation in focusing only on its hypervisible devastation. Studies on audience responses to big-budget cli-fi disaster films seem to show that, indeed, their spectacular nature tends to make climate disaster seem fictional, or at least far away from everyday concerns (Leiserowitz 2004; Reusswig 2005; Lowe et al. 2006; Crespo and Pereira 2013; Svoboda 2015).

If the last few years of dealing with the COVID-19 pandemic has reminded us of anything, it’s in fact that there are many threats and disasters linked with encroaching climate change that are far from easily representable in the form of hypervisible, spectacular event. On the contrary, climate disaster is present in many more insidious ways, in the form of *invisible threats* that have potentially no less devastating effects than their more spectacular counterparts. Indeed, while the coronavirus has been perhaps the most recent of these invisible threats to both exist on such a global scale and to affect the human population in so direct and quantifiable a manner, this idea of understanding impending ecological disaster as being essentially invisible is far from new. In Rachel Carson’s seminal and pioneering 1962 book *Silent Spring*, widely recognized as one of the first texts credited with raising ecological consciousness in the general population, there are potent descriptions of ecological devastation as an *invisible, silent killer*. The first chapter of the book frames the havoc wreaked on the environment by pesticide use as an almost otherworldly, ‘strange stillness’ (Carson, 2002: 2) that takes over rural communities. Carson describes the ecological catastrophe not as spectacular event, but rather as a progressive ‘shadow of death’ (Carson, 2002: 2), a ‘grim spectre’ that has crept upon the citizens of the United States, ‘silenc[ing] the voices of spring’ (Carson, 2002: 3).

Others have more recently pointed out the need to consider, when talking about ecological catastrophe, what Rob Nixon calls 'slow violence', that is, destruction that is not defined as 'an event or action that is immediate in time, explosive and spectacular in space, and as erupting into instant sensational visibility' (Nixon, 2011: 2). For Nixon, we must turn our attention to 'calamities that patiently dispense their devastation while remaining outside our flickering attention spans—and outside the purview of a spectacle-driven corporate media' (Nixon, 2011: 6). Similarly, Timothy Morton speaks of 'hyperobjects', of which global warming is an example, those phenomena that are so 'massively distributed in time and space' that they transcend the local and operate at timescales that entirely exceed human experience, and yet are 'viscous' in that they are inseparable from the (local, immediate) beings that are involved with them (Morton, 2013: 1). Understanding global warming in these terms means, for Morton, that art wanting to engage with environmental issues cannot simply represent the effects of global warming, for 'with objects this huge, this massively distributed, this transdimensional, it's not enough to simply use art as candy coating on top of facts' (Morton, 2013: 182). Rather, writes Morton, hyperobjects call for art that 'walks [people] through an inner space that is hard to traverse' (Morton, 2013: 184).

All three—Carson, Nixon and Morton—conceptualise climate catastrophe not merely as spectacular event, but rather call our attention to its silent, progressive effects, to the ways it creates devastation slowly over time and across space that is not so easily localisable. The spectacular events, indeed, are in these conceptualisations simply the tipping points that explode into hypervisibility exactly *because* we are incapable of attending to the invisible threats as they accumulate around us. Not only this, but it is the very hypervisibility of these events that has a tendency to further hinder our ability to pay attention to those invisible and silent threats before it is too late. The issue becomes, then, how to take into account the non-spectacle, the dangers and forces that aren't so easily translated into the immediately visible event, contained in a precise time and place or easily identifiable body or bodies.

The coronavirus, to a certain extent, or at least the particular way it affects us and spread, is an example of one of these invisible threats, difficult to translate into hypervisible spectacle or easily representable foe. No wonder, on the one hand, that even quite some time after the beginning of the pandemic, there are still those that refuse to believe that it truly exists, that it is something that we should be afraid of, or that we should be modulating our behaviour to combat it: if I cannot see something, how can I believe that it exists? No wonder, also, that there have been so few cinematic representations of our experiences these last few years that seem to adequately describe them.[2] Indeed, the need to address the invisible threats of climate catastrophe poses an issue that is truly cinematic: how can cinema be used to *represent* that which is precisely *unrepresentable*? Or, perhaps more radically, more profoundly, is representation as a paradigm for knowledge *even up to the task* when faced with the problems posed by ecological thought?

Of course, these are vast and difficult questions to answer sufficiently in the confines of this chapter. There are also not many films or audio-visual works that I know of (yet) that address this question in regards specifically to COVID-19. However, I propose here, through analysis of some short experimental films made in the last fifteen years, to examine contemporary film practices that can perhaps serve as models for addressing this question, or, at the very least, that point to interesting ways of thinking through the problem posed above.

Degradation #1: Progressive Erasure

In 2007, the American filmmaker and 'vampler' (video/audio sampler) James June Schneider made two very short experimental films that were part of the same series, titled *Degradation #1: X-Ray, Shroud of Security* and *Degradation #1, X-Ray: Part 2. Government Radiation*. Schneider's filmography over the last thirty years or so can more or less be divided into two categories. Firstly, the films that he is perhaps best known for are his documentaries on punk music (*Blue is Beautiful* [1997], *The Band That Met The Sound Beneath* [2012], *Punk The Capital: Building a Sound Movement* [2019]). The second half of his films tend to principally be explorations of the ways contemporary society creates or destroys certain types of places. His *Dystopian Trilogy*, for example, scrutinises the carceral structure of late capitalist society through the evocation of different spaces: Euro-Disney in Paris (*Faerie-Monition* [1993]), a gated community near Las Vegas (*OASIS* [1995]) and, lastly, in *Median Strip* (1999) highways juxtaposed with prisons. To a certain extent, we can understand the *Degradation* films in continuity with this second half of his filmography (though their sensibility is never far from a certain punk nihilism): they are both evocations of specific organisations of space in contemporary society, and the way invisible forces affect us literally and materially.

In the second of the two *Degradation* films, *Government Radiation,* we see a few minutes of slowed down footage of the U.S. Capitol Building. It's a sunny day, the sky is clear and blue, we can observe a small number of people walking in t-shirts and shorts amongst the trees in front of the national icon. This continuous static shot is interspersed, however, by yellow intertitles on a black background, which simply state a number of 'times': the first is 0 times, then 4, then 8, then 16, 32, 64 and lastly 108 times. After each intertitle, especially after the one stating eight times, the image becomes noticeably foggier, grainier. Incrementally, bit by bit, the image becomes obscured by what seems like a white grainy veil, progressively effacing the image of the building, until at last, in the final image, all that remains visible is a completely washed-out white screen, the Capitol reduced to a gold-tinged, hazy, barely recognisable silhouette.

Watching the film, we as spectators understand how this effect was produced thanks to the opening titles, which tell us, in the same yellow font on black background, that the filmstock (which we're informed is Kodak-7205, 250 ASA) has been 'passed through US government building security checkpoints (X-ray machines) [in] Washington DC'. The progressive erasure of the Capitol from the image is thus a direct result of passing sections of the 16 mm film stock through X-ray security machines various numbers of times. Indeed, Schneider had filmed 100 feet (around 30 metres) of 16 mm footage of the Capitol beforehand, cut it into seven equal parts, then walked through security checkpoints a specific number of times with each piece of film.

Shroud of Security was made in a similar manner, except that in that film, it was one hundred feet of un-shot 16 mm Kodak-7289, 800ASA that was cut into six equal pieces, and then sent to friends of the filmmaker who passed through airport security with each Sect. 0, 4, 8, 16, 32 and 64 times. The accumulative effect of the radiation from the airport X-ray machines on the last section of the filmstock rendered the emulsion almost entirely transparent. The 'shroud' of the title is a reference to the Shroud of Turin, on which the face of Jesus is said to be inscribed. As with the religious relic, *Shroud of Security* purports to have come into contact with, absorb and make visible a local manifestation of some broader, omnipresent figure: not God, but rather both the watchful eye of security infrastructures and the radioactive X-rays that interact with our material bodies.

In both films, there is thus an odd dialectics at work. The viewer sees in these films the progressive erasure of an image, or the emulsion becoming more and more transparent. Both films stage a certain *making absent*—of the figures represented or of the emulsion. However, this move towards absence attests in itself to a particular presence—the presence in the atmosphere of radioactive X-rays, with which the filmstock comes into physical contact, waves that are usually invisible to the naked eye. At the very moment that these waves are made visible (present) thanks to the contact with the emulsion of the film, they seem to make invisible that which was previously there. A dialectic, then, between presence and absence, visibility and invisibility, where one state leads to the other and vice versa during the process of creating the film.

Quiet Zone: Agitated Molecules

The second film we can take as an example is perhaps more empathetic, more human and narrative than the two *Degradation* films. *Quiet Zone* is a short, fifteen-minute experimental documentary made by Karl Lemieux and David Bryant in 2015, which takes as its subject a number of sufferers of what has been named 'electromagnetic hypersensitivity' (EHS), essentially an allergy to being exposed to electromagnetic fields like those created by mobile phones, wi-fi and other electronic devices (Stein and Udasin 2020; Dieudonné 2020). David Bryant is quite well known as an experimental musician, and most famously a part of the Canadian post-rock and multimedia band *Godspeed You! Black Emperor*. Karl Lemieux has also been a part of the band since 2010, providing the film projections at their live concerts. He is also an important figure in the Montreal underground film scene, co-founding the Double Negative collective in 2004 and is a staple of the film lab scene in North America. Both met through working together in the band, and *Quiet Zone* is their first, and for the moment only, film they've made together.

The film itself is made up of interviews that we hear in voice over of two women who suffer from EHS, and who have had to move to the National Radio Quiet Zone (NRQZ) in West Virginia. In this area of around 34,000 square kilometres, radio transmissions and devices that emit electromagnetic waves are severely limited, facilitating scientific research and the gathering of military intelligence from the Green

Bank Observatory. Since the mid-2000s, this area has become somewhat of a gathering place for those who suffer from EHS, creating a small community of 'wave refugees' (Stromberg 2013). The voices that we hear describe what it is like to live with EHS and their decision to move to the NRQZ. They talk of feeling 'exhausted', 'agitated', as if the 'molecules of [their] body' were 'racing around energetically'. They also explain the difficulty of finding themselves forced to be 'refugees' from modern society, the loneliness and the stress that comes from having to find a way to shield themselves from the omnipresence of electromagnetic fields. Their voices are weary yet determined, which gives the documentary a resolutely melancholy tone, amplified by the soundtrack, composed by Bryant.

Bryant's soundtrack also intermittently integrates high-pitched frequencies in order to give an idea of the uncomfortable, agitated feeling that the women describe, but it is especially in through the image-track that the film most obviously attempts to evoke the effects of the electromagnetic sensitivity. We see images of Green Bank and the surrounding landscapes, and some brief moments of who we assume are the women to whom the voices belong. These images are perturbed, however, by various disturbances: flickering stains, what seem like grainy and colourful tears in the fabric of the figurative image, wavy bumps or even what appear to be tumours, jittery colour reversals and solarisations, and pulsating veins that course across the screen. All these effects distort the image on the screen, invading the more or less static and calm shots with a frenetic, beautiful yet ultimately ominous energy. They seem to try and make palpable the suffering that these women feel, caused by the electromagnetic waves, giving the audience a visual and sensorial clue as to the distortion that this particular syndrome creates in the way that they interact with the world. Through the image and the sound, the spectators are pulled into empathising with the specific conditions of the women speaking, we are asked to feel, sensorially, their pain and discomfort. But equally, we can understand the film itself as an exercise in a certain *material* empathy: the effects we see, standing in for the invisible yet palpable electromagnetic waves and the way that they affect the bodies of the film's subjects, are the result of disturbances to the material substrate of the image itself.

This is an important point for understanding the significance of the image distortion that is visible on the screen: it is not the result of digital effects, nor does it come from overlaying the image in post-production or filtering the image as it is captured. On the contrary, all the different ways that the representational image is transformed are the direct results of chemical inference with the filmstock's emulsion. The visible effects are traces of physical interference on the physical autonomy of the original image. The filmstock has been submitted to a specific chemical process called *mordançage*, invented by the French photographer Jean-Pierre Sudre (1921–1997) in the 1960s, based on the etch-bleaching technique elaborated in the 1890s by the German photography pioneer Paul Liesegang (1873–1949). *Mordançage* is a technique whereby a processed film print is immersed in a solution composed of water, copper chloride, glacial acetic acid and hydrogen peroxide for a certain period of time, then redeveloped immediately (James 2015: 816–818). The *mordançage* solution both chemically bleaches the image and softens the gelatine in the emulsion, separating the darker colours from their physical base, and allowing them to be displaced. This gives the impression of veils through the image, of swirling, dislodged image particles. Like the feeling described by one of the women with EHS, it is the actual material component parts of the image that are physically disturbed, warped and moved out of place in the film, producing these jittery, hallucinatory visual effects. In *Quiet Zone*, this process was used in many scenes of the film, with different sections staying in the *mordançage* solution for varying time periods, from 30 seconds to 30 minutes (Lemieux 2015). This renders an already unpredictable process even more impossible to control, meaning that the effects we see are traces not simply of an artistic hand, but of a collaboration with nonhuman, chemical forces that act on their own. Lemieux unleashes a volatile chemical force onto the physical material of the image and then presents the results. This handmade development process thus stands in for the apparently unpredictable effects of electromagnetic radiation on the human bodies.

In both *Quiet Zone* and the *Degradation* films, indeed, the emulsion and the filmstock become metonymies for human bodies: the sometimes violent disturbance or effacement of the recognisable images are supposed to evoke similar effects on the body. In effect, both films propose that we experience the photochemical image as bodies in themselves, that react *bodily* to external stimulation. French film theorist Emmanuel Siety, in his book *Fictions d'image: Essai sur l'attribution de propriétés fictives aux images de films* [*Image Fictions: Essay on the attribution of fictional properties to filmic images*], describes a similar effect in other films, and calls it 'somatisation' (Siety 2009: 62–66). According to Siety, films which employ somatisation establish an apparent relationship of consubstantiality between the image and its subject, participating in a kind of 'empathetic fantasy' (62): the surface of the image itself is made to seemingly empathise corporeally with its subject. Probably the most well-known examples of this in fiction film are the scene in Ingmar Bergman's *Persona* (1966) where the filmstock seems to split apart and burn, mirroring and extending the mental and physical breakdown of Bibi Andersson's character, or the final sequence of *Two-Lane Blacktop* by Monte Hellman (1971) where the image of James Taylor's The Driver burns, exteriorising his burning love for The Girl and despair at having lost her. In these moments, the image is thus posited as a 'sensitive surface' that 'somatises the suffering of the diegetic bodies and souls' (65). Both Bergman and Hellman's films somatise the inner feelings of their characters through a centrifugal movement: the invisible torment that Alma and The Driver are feeling explodes out onto the surface of the image, rendering it visible on the screen. For *Quiet Zone* and the *Degradation* films, however, the movement is rather centripetal: it is the exterior invisible forces that are made to show themselves through their effects on the film-body. In both forms of somatisation, then, the image seems to empathise, take on and render visible the bodily turmoil that is the result of an unseen but real force, whether it be emotional or environmental.

Parasitic Noise

On the one hand, then, *Quiet Zone* and the *Degradation* films posit the photochemical image as material stand-in for the human body, but they also take as their subject non-spectacular, non-visible threats to the autonomy of that image-body. More specifically, these films are about *noise as threat*. In the realm of information theory, starting with Claude Shannon in the 1940s, 'noise' is defined as that which *disturbs* the transmission of information from a sender to a receiver (Shannon and Weaver 1964: 7–8). It is the static on the old televisions that blurred the image. It's the crackle on the phone which means it's difficult to hear the person on the end. It's also, however, if we generalise a bit further, the wind that makes it difficult to hear someone calling you from across a field, or the mist that obscures the horizon from you, or the water that makes it impossible for you to speak to the person you're submerged with, rendering your words unintelligible. That is to say that noise is in fact that which reveals the density of the channel between sender and receiver. As

Shannon shows implicitly in the chapter 'Discrete Noiseless Systems' of his book *The Mathematical Theory of Communication*, there is no channel, or *medium*, without noise (Shannon and Weaver 1964: 36–64).

Michel Serres, in his book *The Parasite*, expands on the notion of noise, showing it to be the structuring element of all relations. All relations, says Serres, inevitably have to contend with some sort of noise, some sort of 'third' that interferes with that relation (Serres 1982: 52). Thus, the relation itself must always configure itself around this third element. Either the relation tries to organise itself in such a way as to actively *repress* or *eliminate* noise, or, on the other hand, it can try and *incorporate* noise into the relationship itself. Either way, the relationship itself must always configure itself *because* of the noise. Noise, for Serres, is thus to communication what a virus or a parasite is to an organism, or society of organisms. We need only look at the ways in which the structures of societal relations have been reconfigured in the last few years to deal with the Coronavirus to see how the responses have indeed fallen into two camps: attempts at absolute elimination by complete overhaul of how we structure social relations, or politics of 'living with' the virus, incorporating its omnipresence into our daily lives. In the films we are analysing here, the noise in question is the electromagnetic waves or X-rays associated with the technological and societal infrastructure in which we live. Made to be invisible, to be ignored, these films show how we are affected by them bit by bit anyway, how they infiltrate our very bodies, like the 'grim spectre' that Rachel Carson describes. Perhaps more importantly, however, not only do they render these forces visible, palpable, in doing so they show the density of the medium in and by which we operate. Lemieux, Bryant and Schneider's films operate as traces of the anthropogenic medium we call modern infrastructure as including, necessarily, these unseen parasites.

This is the opposite of ecological disaster seen as hypervisible event, which translates global warming into a series of spectacular and destructive aberrations. Instead, these films point to the problem being in the very anthropogenic infrastructures we have built and in which we live. It's the invisible, inaudible forces, the noise that we do all we can to suppress, but that is inextricable from the system itself. And the films seem to suggest that the only way we can take this into account is to look and to listen differently: to try and acknowledge the 'murmur of the world' (Lingis 1994: 69–106), or, following Elizabeth Povinelli's more recent reflections on noise, to engage 'otherwise' with it, acknowledging first how it affects us before translating it into easily understandable language or representation (Povinelli 2016: 125–143).

There is, of course, a representational level to these films. We as spectators observe the way that the figurative, recognisable images are effaced or affected by forces that are usually invisible: X-rays or electromagnetic waves. Thus, on the representative level, we are shown a process that is similar to those described in *Silent Spring* or *Slow Violence*. The represented, familiar forms—bodies, landscapes, buildings—are slowly distorted, made unrecognisable, affected in such a way as to make them more and more difficult to distinguish until they eventually disappear entirely, mimicking the slow real-world erosion of that which is represented on screen. However, none of these films rely solely on this representational level of signification to make the invisible threats be felt.

When watching these films for the first time, it is impossible to fully grasp their meaning without a certain level of explanation of how they were made, what process was used to create the images in front of us. For the *Degradation* films, this explanation is given outright in the first moments of the film, laid out for the spectators in yellow writing on a black background. *Quiet Zone* is less explicit, but arguably it is impossible to fully gauge the impact and meaning of the film without the understanding that this is not a digital film but the result of specific post-production chemical operations onto photochemical filmstock. While it is perhaps true that every film is 'a documentary of its own making', as the aphorism attributed to Jacques Rivette states, there is here something that goes a little further. Schneider, Lemieux and Bryant's films explicitly integrate the material process of image formation and deformation into the very act of seeing and understanding the image, thus inviting the spectator to see the image as having a material existence that is affected by that with which it comes into contact. This is not exactly the same operation as *making visible* in the normal, representational, cinematic sense. We as spectators are not observing something through the mediation of the camera 'eye', represented on screen. That would presuppose a specific relation between the pre-established entities that are the viewer, the observed and the mediating tool of vision. That is, the representational paradigm establishes that there is an object or a subject that has existed, has been captured and/or mimicked in some way, and is then made visible in ways that we as observer with specific ways of seeing are able to respond to. Here, the films seem to fold back in on themselves before the representational image can fully stabilise, in the sense that what we as observers see seems rather to be unstable processes of formation and deformation.

Process vs. Representation

In fact, these films seem to propose experiences that are perhaps closer to what neomaterialist philosopher and physicist Karen Barad calls a 'posthumanist performative approach', created specifically in opposition to the 'representationalist' approach. That is, an approach that 'focus[es] inquiry on the practices or performances of representing, as well as on the productive effects of those practices and the conditions for their efficacy' (Barad 2007: 28). For in the context of trying to apprehend cinematographically these invisible threats, viruses, parasites—of which the coronavirus is the latest but will probably not be the last—these materialist films seem to propose the possibility of redefining what we mean by *making or becoming visible*. The representationalist version of this entails, firstly, a stepping back, adjusting our eyes to take in the bigger picture, getting a broader or more precise view thereby bringing into focus that which we could not see before. In this paradigm, the effects of ecological change are necessarily represented as statistical aberrations or as events that don't fit, that are out of place in the picture of what should be. Secondly, since representationalism posits an absolute difference between object in the world and its representation, it can only engage with that which 'already exists', that which has 'already happened'. Thus, the hypervisible event becomes necessarily privileged, since on the one hand it is that which forces us to notice something is definitely awry, and on the other, its effects are eminently representable. But this form of engagement with the world cannot take into account emergent processes, nor, more importantly, the very conditions of our engagement with the world. The 'noise' that constitutes the density of our embodied relationship with the world, through the 'medium' of our infrastructures, the same 'noise' that constitutes a certain invisible threat, creating the conditions for the hypervisible events before they actually happen, is only visible through material engagement with them. The noisy process described in these films is not just representational: it is *material*.

The films thus *bring noise into visibility* not simply by representing it as event. Rather, they point to the possibility of engaging materially, through a certain form of *empathetic performativity*, with the invisible processes that make up the world with which they have come into contact. They invite the spectators to consider the ways processes disrupt that which seems obvious because easily representable. They present the effects that this integration of material process into the formation of meaning has on the production or destruction of a certain form of visibility. This means, in sum, that *Quiet Zone* and the *Degradation* films point at least to the idea that *visibility* is its own specific form of material assemblage, or process of production. In this sense, they perhaps hint at new ways to use media as a tool to engage with the world around us—not by turning it into spectacle, distinguishing between that which is humanly visible and that which is not, but instead by moving towards shedding representationalism as a paradigm. For representationalism seems powerless against the hyperobjects, against the 'silent spring', the slow violence and the noisy parasites that surround us. Instead, these films, and other films of 'process cinema' (MacKenzie and Marchessault 2019), perhaps take a step towards the idea that we must first focus on the *processes* and the *becomings* that make up the world and our relationship to it, on the multiple effects that they produce, and the ways in which this new focus can disrupt our ways of relating to the world. Only then, perhaps, can we attend to the *roots* of the climate catastrophe that is unfolding before our eyes.

Rasheed Araeen, in his 'Ecoaesthetics—A Manifesto for the Twenty-First Century', proclaimed that any ecologically conscious art must abandon the idea of representation and instead focus on the notions of continual transformation and process, only then can art reintegrate itself with 'the collective struggle of life today' (Araeen 2009: 682). It is not the goal of this paper to suggest that Lemieux and Bryant's *Quiet Zone* or Schneider's *Degradation* series are revolutionary works that have the power to ultimately make a real impact on the ecological situation we are faced with. Given the enormity of the problem, it would be foolish to expect to find solutions in the making of art. However, if film can participate in the collective struggle that Araeen speaks of, it is perhaps at the very least by helping us grasp what is happening by pointing to where, and perhaps more importantly *how*, we should look.

Notes

1. "Cli-fi", referring to "climate fiction", was coined in 2007 or 2008 by Taiwan-based American journalist and climate activist Dan Bloom (Glass 2013), referring at first especially to literary works, but

expanding to include all forms of fiction that "use the scientific paradigm of anthropogenic global warming in their world-making" (Andersen 2020: 5).

2. Judd Apatow's last film, *The Bubble* (2022) is perhaps analysable through this lens: its generally agreed-upon failure as satire is perhaps at least in part due to the irreconcilable nature of the spectacularity of Hollywood blockbuster aesthetics and the mundane irrepresentability of the virus and its everyday effects on our lives.

References

Andersen, Gregers (2020), *Climate Fiction and Cultural Analysis,* Abingdon: Routledge.

Apatow, Judd (2022), *The Bubble*, USA: Netflix.

Araeen, Rasheed (2009), 'Eco-aesthetics – A Manifesto for the Twenty-First Century', *Third Text,* 23:5, pp. 679-684.

Barad, Karen (2007), *Meeting the Universe Halfway. Quantum Physics and the Entanglement of Matter and Meaning,* Durham and London: Duke University Press.

Baudrillard, Jean ([1979] 1990), *Seduction*, Montreal: New World Perspectives.

Baudrillard, Jean (1996), 'Illusion/désillusion esthétique', *TRANS,* 7, pp. 59-67.

Carson, Rachel ([1962] 2002), *Silent Spring,* New York: First Mariner Books.

Crespo, Inês and Pereira, Ângela Guimarães (2013), 'Climate change films: fear and agency films', in T. Gustafsson and P. Kaapa (eds), *Transnational Ecocinema: Film Culture in an Era of Ecological Transformation*, Chicago: Intellect Books, pp. 165–186.

Devlin, Dean (2017), *Geostorm*, USA: Warner Bros.

Dieudonné, Maël (2020), 'Electromagnetic hypersensitivity: a critical review of explanatory hypotheses', *Environmental Health,* 19:48, 6 May, https://ehjournal.biomedcentral.com/articles/https://doi.org/10.1186/s12940-020-00602-0. Accessed 15 March 2022.

Emmerich, Roland (2004), *The Day After Tomorrow*, USA: 20th Century Fox.

Fleetwood, Nicole (2011), *Troubling Vision: Performance, Visuality, and Blackness*, Chicago: University of Chicago Press.

Glass, Rodge (2013), 'Global Warning: The Rise of 'Cli-Fi.'' *The Guardian,* 31 May, https://www.theguardian.com/books/2013/may/31/global-warning-rise-cli-fi. Accessed 1 March 2022.

Hellman, Monte (1971), *Two-Lane Blacktop*, USA: Michael Laughlin Enterprises.

Hitt, Christopher (1999), 'Toward an ecological sublime', *New Literary History*, 30:3, pp. 603–633.

Hunnicutt, Gwen (ed.) (2020), *Gender Violence in Ecofeminist Perspective. Intersections of Animal Oppression, Patriarchy and Domination of the Earth*, London and New York: Routledge.

James, Christopher (2015), *The Book of Alternative Photographic Processes*, Boston: Cengage Learning.

Leiserowitz Anthony (2004), 'Before and after *The Day After Tomorrow*: a U.S. study of climate change risk perception', *Environment*, 46, pp. 22–44.

Lemieux, Karl (2015), '*Quiet Zone*. Directed by Karl Lemieux. Interview', *Go Film Magazine*, n.d., http://www.gofilmmagazine.com/2706/quiet-zone/. Accessed 15 March 2022.

Lingis, Alphonso (1994), *The Community of Those Who Have Nothing in Common*, Bloomington and Indianapolis: Indiana University Press.

Lowe, Thomas, Brown, Katrina, Dessai, Suraje, de França Doria, Miguel, Haynes, Kat and Vincent, Katharine (2006), 'Does tomorrow ever come? Disaster narrative and public perceptions of climate change', *Public Understanding of Science*, 15:4, pp. 435-457.

MacKenzie, Scott and Marchessault, Janine (eds.) (2019), *Process Cinema. Handmade Film in the Digital Age*, Montreal and Kingston: McGill-Queen's University Press.

Morton, Timothy (2013), *Hyperobjects. Philosophy and Ecology after the End of the World*, Minneapolis: University of Minnesota Press.

Nixon, Rob (2011), *Slow Violence and the Environmentalism of the Poor*, Cambridge, Massachusetts: Harvard University Press.

Peyton, Brad (2015), *San Andreas*, USA: Warner Bros.

Pidduck, Julianne (2011), 'The Visible and the Sayable: The Moment and Conditions of Hypervisibility'. In F. Grandena and C. Johnston (eds.), *Cinematic Queerness: Gay and Lesbian Hypervisibility in Contemporary Francophone Feature Films*, Oxford: Peter Lang, pp. 9–40.

Povinelli, Elizabeth (2016), *Geontologies. A Requiem to Late Liberalism*, Durham and London: Duke University Press.

Reusswig Fritz (2005), 'The international impact of *The Day After Tomorrow*'. *Environment*, 47, pp. 41–43.

Serres, Michel ([1980] 1982), *The Parasite*, Baltimore: The John Hopkins University Press,

Shannon, Claude and Weaver, Warren (1964), *The Mathematical Theory of Communication*, Urbana: University of Illinois Press.

Siety, Emmanuel (2009), *Fictions d'images, essai sur l'attribution de propriétés fictives aux images de films*, Rennes: Presses Universitaires de Rennes.

Stein, Yael and Udasin, Iris G (2020), 'Electromagnetic hypersensitivity (EHS, microwave syndrome) – Review of mechanisms', *Environmental Research*, 186, 30 March, https://doi.org/10.1016/j.envres.2020.109445. Accessed 15 March 2022.

Stromberg, Joseph (2013), 'Refugees of the Modern World', *Slate,* April 12, https://slate.com/technology/2013/04/green-bank-w-v-where-the-electrosensitive-can-escape-the-modern-world.html. Accessed 15 March 2022.

Svoboda, Michael (2015), 'Cli-fi on the screen(s): patterns in the representations of climate change in fictional films', *Wiley Interdisciplinary Reviews: Climate Change,* 7:1, pp. 43-64.

CHAPTER 10

Noticing Landscape, Sensing Climate: Extending Ecocinema Through Expanded Documentary

Hannah Brasier

Beginnings

In the summer of 2019–20, in the inner-northern suburbs of Naarm (Melbourne, Australia), I woke to the smell of smoke and thought the house was on fire. I was reading David Carlin and Nicole Walker's *The After-Normal* (2019) and MECO Network's *100 Atmospheres* (2019). *The After-Normal* lists 'earthly objects' in alphabetical order, looking for 'joy in what was, what is, and what could be', and in *100 Atmospheres*, 13 writers respond to various ecological ideas from place to oxygen to erosion (Carlin and Walker 2019: xii; MECO Network 2019). Written in list-like collaborative fragments, these creative nonfiction texts evocatively respond to issues of climate change by staying in the present to draw attention towards Anthropocentric damage. This privileging of presence

H. Brasier (✉)
RMIT University, Melbourne, Australia
e-mail: hannah.brasier@rmit.edu.au

D. Binns and R. Najdowski (eds.), *Confronting the Climate Crisis*, Palgrave Studies in Media and Environmental Communication,
https://doi.org/10.1007/978-3-031-89606-4_10

is argued to elicit hope in the face of climate change by anthropologists Anna Tsing and Donna Haraway, who call for 'new tools of noticing' and 'to stay with the trouble', respectively (2015: 25; 2016: 2). For Tsing and Haraway, climate change is a pressing topic we must feel to care for, and this entails noticing in the present to elicit hope for the future.

I am not a writer or anthropologist. Ecological Anthropology has become a lens for me to conceptualize an online interactive audio-visual nonfiction practice concerned with the everyday, travel, landscape, environment, constraint and duration, to attend to the world as soft, multiple, open, indeterminate, precarious, repetitive, excessive and fragmented. Each interaction with my previous films, *Sunny, Rainy, Foggy* and *Sometimes I See Palm Trees* is a dance between nonfiction images of the world, myself as author, Korsakow as the software I use to program interactivity and the viewer turned user because they must interact for the film to continue (Brasier, 2016, 2018b). The experience of these works invites the user to contemplate through their affective interaction. Scott MacDonald describes Ecocinema as a 'depiction of the natural world within a cinematic experience that models patience and mindfulness'. While my films are not strictly cinematic experiences, affective interaction with fragments of landscapes resonates with the 'patience and mindfulness' of the Ecocinema experience MacDonald describes (2013: 19). Can this 'patience and mildfulness' towards landscape of the Ecocinema experience offer a tool for noticing climate change in online audio-visual interactive nonfiction (MacDonald 2013: 19)? By weaving together ecocritical philosophy and Ecocinema film theory I propose a practice of attuned noticing for interactive audio-visual nonfiction which stays in the present to elicit hope towards landscape environments.

The house was not on fire. This smell of smoke drifting in through the evaporative cooling vents marked the beginning of a summer in so-called Australia which saw five and a half million hectares of the east coast burnt (AIDR 2020: n.pag.). Towards the end of this summer, after the fires had stopped, I went for a one-week family holiday to Womboyn Lake on the land of the Bidawal people. Womboyn Lake can only be accessed by one road, and as we drove in, every noticeable tree was burnt, yet at the end of this road, the town, which consists of a caravan park and scattered houses, seemed to have come away largely unscathed. In the conclusion of *Attunings,* I note how 'aesthetic detail' is 'only one of the textures' caught by my previous interactive audio-visual nonfiction and I see 'potential for other rhythms to be seen as well, perhaps in Tsing's "damaged landscapes"' (Tsing cited in Brasier 2018a: 106). *There's Still Heaps of Water* (Brasier, 2022) takes up this question to consider what rhythms emerge in the aftermath of 2020's Black Summer in the damaged landscape of Womboyn Lake.

Damaged Landscape

When I was little I visited Womboyn for fishing trips with family friends. We would cover ourselves in bird seed, hoping a rainbow lorikeet, or even better, a king parrot would land on our shoulders. This time I am confronted by diseased animals, notably the insides of a whale washed up on the beach. As we drive around, I notice a tension between the aesthetics of the landscape and the devastation of the land. There is something eerily beautiful about the blackened landscape, yet the immensity of what has been damaged by these fires feels heavy. In conversations, 'I really don't want to see the burnt houses, that's a bit depressing' is said in almost the same sentence as the leftover stumps and bolts of a burnt jetty are 'fantastic for photos'. We heard the bushfires had devastated the area before we left Naarm, yet somehow there was an expectation that our summer holiday would be the same as every other. Many of our usual places to go were at the end of roads closed due to risks of tree branches falling. While my dad noted Womboyn Lake 'had heaps of water', the forest surrounding was largely burnt, the beaches were covered in ash and a few of the houses were haphazard piles of wood wrapped in orange caution tape.

Filming for *There's Still Heaps of Water* became about noticing this aftermath, recalling Tsing's question, 'what emerges in damaged landscapes' beyond stories of 'promise and ruin' (2015: 18)? As a response, Tsing offers a practice that attends to 'what's left'?; a consideration of what lies beyond or outside of these linear narratives of 'promise and ruin' (2015: 18). Black Summer was marked by news stories of fires storming along the eastern coastline as residents fled into the water to escape the flames, later to be rescued by a navy ship (Hennessey 2021: 27). Tsing notes how such devastating stories are what '*we need to know*' to empathize with environmental damage, although simultaneously we need to sit within these landscapes longer to feel their weight (2015: 18, original emphasis). Haraway describes this lingering in the presence as 'staying with the trouble' which 'requires learning to be truly present, not as a vanishing pivot between awful or edenic pasts and apocalyptic or salvific futures, but as mortal critters entwined in myriad unfinished configurations of places, times, matters, meanings' (2016: 1). *There's Still Heaps of Water* lingers within the aftermath of the bushfires in an attempt to 'stay with the trouble' (Haraway 2016: 2). As I walk around the outskirts of the lake with my phone as a camera, I think about filming what is left, where turning the landscape into an aesthetic experience troubles me somewhat. I am encouraged by Tsing's words: '*if we end the story with decay, we abandon all hope*' (2015: 18, original emphasis). *There's Still Heaps of Water,* in its aesthetic portrayal of the Womboyn Lake landscape, aims to move beyond decay to allow the user to see hope in what remains.

The discomfort I felt in turning the devastation of the fires into an aesthetic experience for an audience reflects upon the relationship between aesthetics and ethics in documentary filmmaking. Noted in Patricia Aufderheide, Peter Jaszi and Mridu Chandra's *Honest Truth's* report, documentarians have 'three conflicting sets of responsibilities: to their subjects, their viewers, and their own artistic vision and production exigencies' (2009: 1). The aesthetic choices of a documentary filmmaker always have an ethical dimension as selecting what to film, framing, camera movement and the process of omitting footage in the edit 'treats actuality' according to a range of personal, industrial and technical decisions and constraints (Grierson 1933: 8). For *There's Still Heaps of Water* multiple facets of damage within the Womboyn Lake landscape was my subject, so I considered how I could surrender some of my artistic vision as a filmmaker to the landscape itself. How could the landscape partly determine what was filmed? In the Ecocritical Anthropology of Jane Bennett and Tsing, they speak of how landscapes 'shimmer' when their multiple entities are not reduced 'to the contexts in which (human) subjects set them' (Bennett 2010: 5). They note how these entities are continually shifting and evolving as 'they gather rhythms shared across varied forms of life' (Tsing et al. 2017: G12). 'Shimmer' suggests entities of landscapes can call attention towards themselves if we have the 'anticipatory readiness' necessary not to condemn them to our preconceived notions but to recognize them as constantly shifting in their entanglements with others (Bennett 2010: 5). As I walk around the lake, I capture what I notice shimmering in response to what is left. The posts, bolts and screws as the only remnants of a boardwalk curving lines through swampland, the carcass of a kangaroo, reeds swaying to- and fro-, ash blowing gently across the sand and the water ripples reflections of pink at sunset.

Attuned Noticing and Technology

Tsing's call for 'new tools of noticing' 'what's left' requires a reframing in my online interactive audio-visual practice where the ecology of noticing expands beyond myself to include my phone's camera, the software I use to program interactivity and the user (2015: 25, 20). While noticing makes a 'distinction to create foreground and background, to distinguish some "thing" from its surroundings' (Mason 2002: 33, original emphasis), attunement 'invokes a particular sensitivity—receptive and curious, careful and attentive. A deliberately hesitant engagement with other entities on their own terms' (Kuzmanovic et al. 2019: 3). In an audio-visual context, Silke Panse argues that documentary often starts from a preconceived idea that funnels what is filmed through the lens of human perspective. She notes 'there are not many films which actually *look*' (Panse 2013a: 66, original emphasis). Together, John Mason along with Maya Kuzmanovic, Nik Gaffney, Ron Broglio, Adam Nocek and Silke Panse provide an approach to audio-visual nonfiction that centres the landscape as a human and nonhuman environment. Attuned noticing requires a mode of looking that distinguishes individual things to pay attention to their multiple forms while following the complex web of human and nonhuman entanglements they are part of in the landscape.

My approach to *There's Still Heaps of Water* was an attempt to centre the landscape through this practice of attuned noticing. To begin this centring, I curiously walked around, stopped and took in what was surrounding me without a preconceived notion of what I would film or what the film would be about. Here, I let what made me stop suggest what was to be filmed. To distinguish individual things on their own terms, I provided myself with constraints external to my subjective position as a filmmaker; each shot was limited to ten seconds with no camera movement. By imposing these constraints, I allowed the indifference of my phone's camera to notice what was in front of it without providing myself the ability to leave the camera running until something happened or moving my camera to follow action. Additionally, to follow the complex web of human and nonhuman entanglements a particular thing was part of, I would research and write lists of what things were connected to and use these lists as a prompt for further filming. As I was following this process of distinguishing and connecting elements as a practice of attuned noticing in the landscape, I realized areas of improvement in centring the landscape. Largely, I was still selecting and framing shots, disallowing me to negate my role as a filmmaker to the indifference of my camera. Avoiding a preconceived idea and filming according to external constraints provides two methods of capturing nonfiction content which distributes the weight of decisions to allow patterns to emerge between myself, landscape and a phone's camera. Lessening my subjective control of the image by providing my camera agency in what it notices moves me closer to privileging the interconnected human and nonhuman entanglements of a landscape on their own terms.

This distributed practice of filming landscape from a perspective which 'actually *look*[s]' is seen in the landscape films of James Benning (Panse 2013a: 66, original emphasis). Panse considers Benning's films to *look* because he goes to a location, sets up his camera and allows it to capture what it sees, rather than relying on pre-established shots. In his films, duration is often pre-determined by the length of a piece of film or an arbitrary time frame, such as how long it takes someone to smoke a cigarette. In *Twenty Cigarettes*, the person smoking presses record when they light their cigarette, sits back in front of the camera and stops the recording when they have finished (Benning, 2011a, b: n.pag.). While *Twenty Cigarettes* is not one of Benning's landscape films, these arbitrary constraints provide a tool to 'actually *look*' because the camera's indifference to what it notices is privileged over what the filmmaker is drawn to film (Panse 2013a: 66, original emphasis). Panse's theorization of Benning's films shows it is possible to come closer to capturing what is out there as opposed to filmmakers imposing ideas and agendas on the world. By applying arbitrary constraints, a camera's indifference has agency to notice what it sees and, in doing so, becomes a technological nonhuman 'new tool for noticing' in an audio-visual nonfiction practice (Tsing 2015: 18).

Online Interactivity and the Balancing of Agency

In addition to beginning without a preconceived idea and applying external constraints, the online interactivity afforded by the Korsakow software, which I use to program my films, further extends the distribution of agency. In my wanderings around Womboyn Lake, I collected 179 10-second videos, which I imported into Korsakow as individual self-contained clips. There is no linear timeline in Korsakow, instead, each clip remains individual and a system similar to tagging is used to create unfixed associations between videos. What a user sees when interacting with a Korsakow film (K-film) on a computer is a playing video and a set of thumbnails that provide options for the viewer on what video to open and play next. In *There's Still Heaps of Water*, this interface shows a playing video followed by three identically sized thumbnails displayed horizontally across the middle of a white screen. The three thumbnails show stills of the videos and play when the user hovers the cursor over them. The tags I have used to associate the video fragments are hidden from the user to encourage them to notice certain resonances across the clips without prescribing that they will notice these things. This indeterminacy of how the user will navigate the work extends the distribution of agency in Korsakow filmmaking.

To determine what tags I will attribute to my clips, I use Korsakow as a tool to attune to patterns emerging in the footage, furthering my attempts to follow human and nonhuman entanglements. I import ten clips into the project, export a random K-film and use a practice of listing to record the associations that I notice forming through the software's agency. Buzz, revving, crashing, tweeting, surfaces, horizons, stone, twirling, SunSmart, 'it's a bit wild with that breeze behind us' and trickling provide examples of some of the first things I list as I view and listen to my clips through a random export in Korsakow. Similar to providing my camera's indifference agency while filming, I allow Korsakow's indifference to what clips it selects to associate with each other for me to notice resonances, patterns, themes and differences. I developed this practice of listing to notice through *Attunings*, where the 'open, flat, indiscriminate, interested, and partial' qualities of a list became a way of attending to what goes unnoticed (Miles 2016: n.pag.). I use this practice in *There's Still Heaps of Water* to draw attention to unnoticed qualities within my video fragments.

The listed resonances of surface, landscape and movement emerge as the tags I use through repeating this process of listing to notice. Surface and landscape loosely organize the film into a pattern of zooming in and out of the damages caused by the bushfires while movement draws attention to travels through the environment by car and boat. This noticing with Korsakow can be considered a dance of interactivity, where the way that Korsakow associates clips across the interface has agency in deciding the shape of the film and the user's experience of the work. A user notices, clicks and then the software provides a new set of video thumbnails for them to select from based on my loose structuring. Agency is then distributed from the software to the user, whose decisions on what video to click next shape a different version of the film through every interaction. By distributing agency more evenly between a shimmering landscape, software and user, *There's Still Heaps of Water* moves closer towards negating the ethical dilemma of framing the damaged landscape of Womboyn Lake according to my own aesthetic preferences, and in doing so privileges the micro and macro aesthetics of the landscape.

Beyond Story in Expanded Documentary Practices

This diminishing of authorial control to privilege the damaged landscape of Womboyn Lake contributes to a recent wave in documentary film scholarship which questions why story dominates the practice. These film theorists call for alternative ways of organizing nonfiction content to reflect the complexity of the world (Gustafson and Lusztig 2019; Juhasz and Lebow 2018, 2021; Miles 2016). Alexandra Juhasz and Alisa Lebow argue that the 'constricting contours' of story are insufficient in capturing experiences that are 'unmoored, confusing, unstorified, especially when experienced in crisis' (2018: n.pag.). Adrian Miles suggests stories make 'too neat something that was intrinsically messy, ambiguous, dangerous, and multiplicitous' (2016: n.pag.). As stories make sense of the world according to openings, closings and cause-and-effect logic, they cannot possibly capture the complexities of an actuality which functions in excess of such a neat structure. Working with Korsakow, through a practice of listing to notice, allows organizing of nonfiction content which negates locked-off linear relations between one shot and another and therefore maintains each nonfiction bit of the world as open to multiple relations. Womboyn Lake is a complex ecosystem which exceeds my understanding of it, so *There's Still Heaps of Water* does not attempt to make sense of this landscape according to a fixed linear trajectory.

If a story about the relations of Womboyn Lake is negated in *There's Still Heaps of Water*, what does an interactive audio-visual nonfiction project which draws attention to qualities of climate change do? Samantha Lang's *Brown Lake* (2021) and Isabelle Carbonell's *The Mississippi Multiverse* (2021b) and *The River Runs Red* (2018) provide three examples of expanded experimental documentaries which use numerous capturing technologies to sense qualities of climate change in specific landscapes. These films 'sense climates' because their approaches can be seen to reflect T. J. Demos, Emily E. Scott and Subhankar Banerjee's two-fold approach to sensing as 'human and more-than-human sensory capacities [...]

defined within and through divergent natural systems, cultural practices, Indigenous traditional ecological knowledge, scientific methodologies, and developing technologies'. In other words, these short, expanded documentaries combine human and more-than-human noticing alongside multi-layered ways of knowing a landscape to allow the audience to 'feel, perceive, detect, apprehend, [and] grasp with the senses' qualities of climate change (Demos et al. 2021: 149).

The landscape documented in Lang's *Brown Lake* is the Bummiera Lake (Brown Lake) on Stradbroke Island, located off the east coast of so-called Queensland in central-north Australia. The 15-minute short reveals the lake through multiple capturing technologies, including aerial photographs which zoom out from the micro swirls of the lake's surface to macro aerials which situate the lake geographically to reveal industrial development. Abstract virtual reality representations show viewers the curving lake bank as a voice-over situates the lake within Indigenous history. Extreme low angles look through the grasses surrounding Bummiera Lake, inviting audiences to see the landscape from the perspective of an insect emphasized through a textured soundscape of buzzing cicadas. The film culminates with a poem by Indigenous Stradbroke Island poet, Oodgeroo Noonuccal, spoken in native tongue without subtitles, by his granddaughter Kaleenah Edwards (Harris 2021: n.pag.). The combination of multi-layered soundscapes, photography, virtual reality, close-ups, voice-over and Indigenous historical accounts offer the Bummiera Lake as a landscape to be felt from multiple human and nonhuman perspectives.

Carbonell's *The Mississippi Multiverse* takes a similar approach by combining multiple technological 'tools for noticing' with divergent ways of knowing the landscape to sense the Mississippi River (Tsing 2015: 18). In the 26-minute short, GoPro footage looks up at the tree-lined sky from underneath the turbulent wavy surface as a hydrophone picks up the whooshing sounds of the water. Extreme close-ups are used to notice a colony of ants marching precariously over a tree branch. We faintly hear a boat radio captain's reading of the riverscape. Carbonell's use of GoPro footage allows a surrendering of control to the water as the camera captures what it sees from this perspective: blurred trees at sunset, seaweed, flashes of light and a dizzying sky. Further, as the GoPro is directly placed in the water, it is moved by the waves so there is a direct sensorial relationship between camera and water. This entangled relationship between GoPro and the world is discussed by Andrew

Murphie, who considers GoPro aesthetics 'so in the middle of things human and non-human, that they decentre the 'human' [...] The camera is more immediately moved with the forces of the world' (2014: 194). Carbonell describes this multi-faceted approach to documenting the Mississippi River as 'using experimental film methods to attune to these nonhuman stories [;] [...] a type of embodied sensorial practice that focuses on encounters between humans and nonhumans' (2021b: n.pag.). In *The Mississippi Multiverse*, Carbonell uses various capturing technologies to privilege the diversity of the multi-species human and nonhuman occupants of the Mississippi River.

Aligned with *There's Still Heaps of Water* due to its interactivity is Carbonell's *The River Runs Red* which is a 'constellation of films arranged in a poetic multi-linear dialectic' to draw attention to and then sense the damage caused by the iron-ore tailing dam collapsing on the Rio Dolce river and releasing 'chemical-laden sludge' in Brazil, in 2015 (Carbonell 2021a: 146, 140). This 'constellation' consists of 43 films that reveal human and nonhuman responses to the devastation through various technological and filmic interventions. Drones sweep over the river landscape as we hear thundering sound effects, GoPros are attached to humans and mules to attune to the destruction from different perspectives and conventional documentary interviews are conducted to reveal how the collapse has affected the everyday lives of those living in the area. The interactivity of the film is poetic because 'opposing concepts, ideas, words or worlds' such as land/ghosts, dystopia/nostalgia and flood/fauna organize the work as opposed to cause-and-effect logic (Carbonell 2021a: 146). Users interact via a computer's interface by clicking a side in these opposing ideas that appear as text links at the end of each film and once we reach the end of a pathway, text on screen asks questions reaching beyond the documentary's scope. An ending I reach poses 'what other morphologies has this disaster reconfigured in this landscape?' (Carbonell, 2018). We are then invited to enter the project again to take a different pathway, where there are 'many possible combinations of the material, and a viewer may never watch all 43 films/dialectics' (Carbonell 2021a: 146). Using drone, GoPro, interview, poetic interactivity and further questions, *The River Runs Red* goes 'beyond story' to sense the devastation from multiple human and nonhuman occupiers of the land (Juhasz and Lebow 2018: n.pag.).

In comparing *There's Still Heaps of Water* to these expanded documentaries, I now recognize a limitation in my methodology. Upon reflection, the use of additional capturing technologies could have expanded the sensorial capacity of the project. However, *There's Still Heaps of Water* does use what Camille Baker, Max Schleser and Kasia Molger describe as the 'intimate' and 'immediate' affordances of my smartphone's camera to focus on a body of water and its surrounds in largely fragmented close-ups, encouraging noticing towards the unnoticed qualities of how the landscape was burnt (2009: 101). These intimate close-ups include the burnt orange tips of leaves swaying in the wind, piles of ash on the beach, a fern sprouting up in largely burnt surroundings and the lake rippling at sunset. These fragments of surfaces are interspersed with wide shots to situate these details within the larger environment of Womboyn Lake. There are numerous shots filmed from a car moving along the highway to signify my position as a tourist and draw attention to the widespread damage of the bushfires. By moving from micro to macro, *There's Still Heaps of Water* invites the user to zoom in and out of the multi-faceted ways the landscape had been damaged by the Black Summer bushfires.

The video fragments of *There's Still Heaps of Water* go further to 'evade story' and sense climate because they are provided for the user to interact with through a process theorized by Miles as 'noticing, deciding and doing' (Miles 2014: n.pag.). Miles explains that in cinema this movement from noticing to deciding to doing is 'overcome [...] through montage which corrals these varieties of images into relations that become fixed in their order and occurrence upon the screen'. In interactive audio-visual nonfiction, there is 'always this ongoing site of indeterminacy located in the user who necessarily becomes an affective relay between perception and action, watching and clicking' (Miles 2014: n.pag.). As demonstrated in *There's Still Heaps of Water* and Carbonell's *The River Runs Red*, the user is required to click a video or text thumbnail respectively for their experience of the work to continue, and therefore the user becomes the affective centre of the work who uses their own contemplative noticing to think through which video to open and view next. This affective space afforded by interactivity can generate poetic nonfiction engagement, where Sable Elyse Smith notes 'each line of a poem is creating an image and there is a reverberation in the space in between each of those lines that are building up to a ghosting image [...]: the space that is in between an image that I might give you and a type of reference that you might locate' (Smith et al. 2021: 115). We can then consider the indeterminate space which opens in *There's Still Heaps of Water* as allowing the user to poetically engage with each video because their position as an affective centre allows them to locate a reference and use that to think through their navigation with the work.

Ecocinema for Online Audiences

This affective noticing can occur when watching Ecocinema films, where extended duration is a predominant technique that allows such an experience. Benning's landscape films show us wide expansive shots of landscapes for durations ranging from two to ten minutes. Benning's *13 Lakes* (2004) provides an example where thirteen identically framed shots of lakes around America last ten minutes each. Every shot is framed with a horizontal split across the horizon between sky and lake while atmospheric sound captures gunshots and laughter drawing attention to what is beyond the visual frame. The duration of each shot invites the viewer to notice changes, such as patterns of fog along the surface, cars across a bridge and ebbs and flows of waves. In my experience of watching *13 Lakes,* I found myself questioning where my attention was being drawn to, where it was not and what I was missing if I focused on one part of the frame or a sound for too long. Benning describes his 40-year long career of watching landscapes as developing a patience to 'see something new happening over time' (2018). This patience to see something shifting over time is encouraged through the extended shots of lakes in *13 Lakes*, whereby leaving a landscape on screen for ten minutes instills a sense of patience and attention towards subtle changes within the frame; to see shifts over time unavailable if such a shot had been fleeting. By deeming 'landscape as a protagonist' through long durational shots, Ecocinema positions the environmental landscape as 'worthy of our sustained attention' (Panse 2013a: 61; MacDonald 2013: 21).

In comparison to the 10-minute shots in *13 Lakes,* the 10-second shots in *There's Still Heaps of Water* are relatively short. However, patient attentiveness is still encouraged because each playing clip loops infinitely until the user selects a new video to play from the thumbnails. This infinite looping invites the user to continually notice additional qualities of a landscape within a shot. In a clip of pink clouds at sunset, the user may not notice the swaying of trees along the bottom or the swaying of powerlines at the top of the frame until the clip has looped multiple times (Fig. 10.1). In my own experience, it was only after twenty or so viewings that I noticed a fishing rod cast out from a shot of a boat moving across the lake (Fig. 10.2). Additionally, each of these looping shots can repeat infinitely in the film, where each clip can reappear as a choice in the thumbnails for the user to view again and again. This allows the viewer to watch each landscape in different relations with other surface shots and vice versa, extending their attunement to discrete details and varying entanglements. For instance, if a landscape shot of a person walking along the beach is playing and the thumbnails show three beach surfaces of seaweed, ash with leaves and ash on the sand, then the user is encouraged to zoom into the discrete details of how the bushfires have affected the sand (Fig. 10.3). If the same beach landscape is selected again from the thumbnails and is associated with the surfaces of green growth, burnt leaves and rain puddling, then the user's attention may be drawn to less apparent qualities such as the movement of water or the forest in the distance (Fig. 10.4). *There's Still Heaps of Water* encourages Ecocinema's attentiveness to landscape through its looping and repetitive qualities, evolving as an Ecocinema experience for online audiences.

The experience of *There's Still Heaps of Water* further resonates with the frequent use of still shots in Ecocinema, which we see in Benning's first digitally produced film, *Ruhr* (2009). The lack of camera movement in *Ruhr* is discussed in relation to an 18-minute long-take by Panse in '*Ten Skies*, *13 Lakes*, 15 Pools' (2013b). This wide extended shot is filled with leafy trees, where repeatedly throughout we hear a plane moving overhead which causes leaves to move from complete stillness, although the movement is delayed until after the sound of the plane is no longer heard. The shot, therefore, can be described as showing an alternating pattern from stillness to movement in response to the plane, where the complete stillness of leaves is noteworthy as it appears as though the shot is completely frozen. If *Ruhr* was shot on film, we would still see the movement of the film grain. Panse suggests this complete stillness of the foliage invites an alternative spectatorial experience because:

> we think moving images should be seen to move, even if the objects depicted in them are momentarily still. The temporary suspension of movement makes us aware of an anxiety that the movement could stop, as if the ceasing of movement in a film means the end of our life that is so determined by images. (2013b: 47)

In other words, stillness in moving images subverts viewer expectations and encourages a patient spectatorial position due to the expectation of

Fig. 10.1 Pink clouds at sunset with trees and powerlines swaying at the bottom and top of frame, respectively, Hannah Brasier (dir.), *There's Still Heaps of Water*, 2022. Australia © Hannah Brasier

movement. Stillness lasting multiple minutes becomes, if not uncomfortable, then at least peculiar for viewers. This uncomfortable positioning invites the viewer to search the image for what is there, rather than draw attention to action that is happening. By avoiding a dramatization of the image, still shots evolve as a 'tool for noticing' Anthropocentric damage because they confront us to see the aesthetics of the landscape in the present moment (Tsing 2015: 18).

Fig. 10.2 A fishing line cast out from a boat into Womboyn lake with burnt trees lining the sandy bank, Hannah Brasier (dir.), *There's Still Heaps of Water*, 2022. Australia © Hannah Brasier

In *There's Still Heaps of Water*, this privileging of stillness can be seen to operate in conjunction with the loop, which offers extended duration in online contexts. As noted previously, the only shots in motion in the K-film are those that are diegetically moved when my camera was in a moving vehicle, so the majority of the clips within the film are still. What results through the inclusion of these shots in motion is a contrast between stillness and movement, which is further emphasized by the

Fig. 10.3 Interface showing a playing clip of a person walking along a beach followed by the thumbnails: (1) seaweed on the sand, (2) ash amongst fallen dried leaves and (3) ash blowing along the sand, Hannah Brasier (dir.), *There's Still Heaps of Water*, 2022. Australia © Hannah Brasier

Fig. 10.4 Interface showing a clip of a person walking along a beach followed by the thumbnails: (1) green leaf sprouting from the ground, (2) burnt orange leaves and (3) rain puddling on the ground, Hannah Brasier (dir.), *There's Still Heaps of Water*, 2022. Australia © Hannah Brasier

user's interaction from contemplative stillness to clicking a thumbnail. As the user can activate thumbnails by scrolling over them, this invites attention towards patterns of stillness and movement. In the process of making *There's Still Heaps of Water,* I noticed such patterns in how a boat moved across the water between the playing clip and the thumbnail next to it.

This created a tunnelling effect between the two clips across the interface design (Fig. 10.5). While movement and stillness are not used as tags to structure the K-film, the user is likely to notice juxtapositions between movement and stillness through their activation of thumbnails.

MacDonald's theorization of Ecocinema then provides a productive lens to consider *There's Still Heaps of Water* as a K-film which asks the audience to slow down their interactions with online media to contemplate the damaged landscape of Womboyn Lake. This invitation is evidenced in the use of looping videos which show details of what is left in the aftermath of the fires, without a narration or story leading the user definitively from one video to the next. By offering video fragments showing ripples on the surface of water, close-ups of blackened tree stumps and movements through the surrounding environment, the user makes their own associations and rhythms through

Fig. 10.5 A tunnelling effect created by two clips aligning next to each other of a boat moving across Womboyn Lake towards the horizon, Hannah Brasier (dir.), *There's Still Heaps of Water*, 2022. Australia © Hannah Brasier

the K-film as a contemplative and affective dance. Matt Soar's comprehensive essay on Korsakow filmmaking echoes MacDonald's theorization of an Ecocinema experience by suggesting K-films require 'attention' and a 'reasonable investment in time without distraction' (2014: 160). He considers K-films an alternative 'slow web' to fast interactions which commonly take place as we flick through content with multiple tabs open. As a result, K-films are 'if not "languid", exactly, then contemplative, interpretative and exploratory: extended spatial montages with minimal interface elements, and certainly no "all-singing-all-dancing", technologically precocious affairs' (Soar 2014: 160, original emphasis). There are obvious parallels here between Ecocinema's '"testing viewers" patience in order to reinvigorate their perceptual capacities' and the 'attention [...] without distraction' encouraged by a K-film (Benning 2007: n.pag., original emphasis; Soar 2014: 160). *There's Still Heaps of Water* encourages affective interaction, which mimics the patience required in viewing Ecocinema. The K-film asks the audience to slow down and contemplate Anthropocentric damage at Womboyn Lake.

Endings

Peter Hutton notes the 'experience [of Ecocinema] offers "a reprieve", from the busyness of our consciousness' (cited in MacDonald 2013: 40). Although, whether we have the attention to absorb these extended shots of landscapes is another question altogether. Yes, we are provided with a still shot of a landscape to look at for an extended period, yet we can become distracted and move unconsciously away from the discrete details of these films. The interactivity required to experience a K-film offers a way to navigate this dilemma as the work directly involves the user moving the film forward by clicking a thumbnail. *There's Still Heaps of Water* can therefore be conceptualized as a dance between author, user, images and software with the caveat that this dance relies upon attunement; 'a particular sensitivity—recessive and curious, careful and attentive. A deliberately hesitant engagement with other entities on their own terms [...] A kind of "ecological intimacy"' (Kuzmanovic et al. 2019: 3, original emphasis). Attunement as 'ecological intimacy' is central to the success of *There's Still Heaps of Water* because the work encourages an attentiveness that is curious to the multi-layered human and nonhuman entities of Womboyn Lake. While the K-film does not use the range of capturing technologies that *Brown Lake, The Mississippi Multiverse* and *The River Runs Red* do, the extensive use of close-ups draws attention towards discrete details of nonhuman interactions while affective interactivity encourages a 'deliberately hesitant engagement' with the landscape (Kuzmanovic et al. 2019: 3).

By starting without a pre-conceived idea, I missed the opportunity for *There's Still Heaps of Water* to combine multiple capturing technologies with interactivity as *The River Runs Red* does; to provide an embodied multi-species sensorial experience of a climate disaster. However, what *There's Still Heaps of Water* does do is provide extended glimpses into the various micro ways bark, sand, ground, water, roads, signs and logs are affected by bushfires through close-ups and the macro ways entire landscapes are affected through wide, and sometimes moving, shots of the forest, lake and ocean. In this way, *There's Still Heaps of Water* contributes to the list-like fragmented attunements to landscape experienced as I read *The After-Normal* and *100 Atmospheres.* My next project would move towards the multi-species sensorial experiences of *Brown Lake, Mississippi Multiverse* and *The River Runs Red* by using numerous capturing technologies and including a multiplicity of knowledges of the land. These additions will expand upon the sensorial capacity of Korsakow as a 'tool for noticing' (Tsing 2015: 18). Here, I can picture a project which responds to Juhasz and Lebow's 'Beyond story' project which demands 'documentary to think beyond the human to include structural analysis and non-human perspectives' as 'the world itself [...] ruptures, sometimes unexpectedly. Documentary is equipped to be with us, exposing crisis, suggesting alternatives, adapting to and with the world it allows us to see' (2021: n.pag.). We live in a highly precarious time, where more than ever documentary has an obligation to reflect and change according to this uncertainty. By expanding an Ecocinema experience through affective interactivity, landscape films authored using Korsakow invite attentive care towards Anthropocentric damage.

References

Aufderheide, Patricia, Jaszi, Peter and Chandra, Mridu (2009), *Honest Truths: Documentary Filmmakers on Ethical Challenges in Their Work*, Washington: Center for Media and Social Impact (CMSI). https://cmsimpact.org/resource/honest-truths-documentary-filmmakers-on-ethical-challenges-in-their-work/. Accessed 22 July 2021.

Australian Institute for Disaster Resilience (AIDR) (2020), *Bushfires - Black Summer*, AIDR, https://knowledge.aidr.org.au/resources/black-summer-bushfires-nsw-2019-20/. Accessed 7 October 2021.

Baker, Camille, Schleser, Max and Molga, Kasia (2009), 'Aesthetics of mobile media art', *Journal of Media Practice*, 10(2–3), pp. 101–122.

Bennett, Jane (2010), *Vibrant Matter: A Political Ecology of Things*, Durham: Duke University Press.

Benning, James (2004), *13 Lakes*, USA: CalArts.

Benning, James (2007), Interviewed by Scott MacDonald, 'Testing your patience: An interview with James Benning', *Artforum*, 46:1, https://www.artforum.com/print/200707/testing-your-patience-an-interview-with-james-benning-15707. Accessed 10 June 2021.

Benning, James (2009) *Ruhr*, DVD, USA: Edition Film Museum.

Benning, James (2011a), Interviewed by Mark Peranson, 'Berlinale 2011: Twenty questions for James Benning on *Twenty Cigarettes*', *Cinema Scope Online*, http://cinema-scope.com/cinema-scope-online/berlinale-2011-twenty-questions-for-james-benning-on-twenty-cigarettes/. Accessed 27 January 2017.

Benning, James (2011b) *Twenty Cigarettes*, USA: CalArts.

Benning, James (2018), Interviewed by Arterritory, *Arterritory*, https://vimeo.com/282216582. Accessed 7 October 2021.

Brasier, Hannah (2016), *Sunny, Rainy, Foggy*, Australia: Korsakow, http://projects.hannahbrasier.com/sunnyrainyfoggy.

Brasier, Hannah (2018a), 'Attunings: Multilinear ways of thinking about, making with, and sensing the world', Ph.D. thesis, Melbourne: RMIT University.

Brasier, Hannah (2018b) *Sometimes I See Palm Trees*, Australia: Korsakow, http://projects.hannahbrasier.com/sometimespalmtrees.

Brasier, Hannah (2022) *There's Still Heaps of Water*, Australia: Korsakow, http://projects.hannahbrasier.com/theres-still-heaps-of-water.

Carbonell, Isabelle (2018), *The River Runs Red*, USA: Klynt, http://www.theriverrunsred.com/. Accessed 22 July 2021.

Carbonell, Isabelle (2021a) 'Multispecies cinema in wretched waters: The slow violence of the Rio Doce disaster', in T.J. Demos, E. E. Scott and S. Banerjee (eds), *The Routledge Companion to Contemporary Art, Visual Culture, and Climate Change*, New York: Routledge, pp. 139–148.

Carbonell, Isabelle (2021b) *The Mississippi Multiverse*, USA: Anthropocene Curriculum, https://www.anthropocene-curriculum.org//contribution/the-mississippi-multiverse. Accessed 6 May 2021.

Carlin, David, and Walker, Nicole (2019), *The After-Normal: Brief Alphabetical Essays on a Changing Planet*, Brookline, MA: Rose Metal Press.

Demos, T.J., Scott, Emily E., and Banerjee, Subhankar (eds) (2021), 'Sensing climates', in *The Routledge Companion to Contemporary Art, Visual Culture, and Climate Change*. New York: Routledge, pp. 149–152.

Grierson, John (1933), 'The documentary producer', *Cinema Quarterly*, 2:1, pp. 7–9.

Gustafson, Irene and Lusztig, Irene (2019), 'Poetics & politics: Against story', In Media Res: A Media Commons Project, http://mediacommons.org/imr/content/poetics-politics-against-story. Accessed 22 July 2021.

Haraway, Donna (2016), *Staying with the Trouble: Making Kin in the Chthulucene*, Durham: Duke University Press.

Harris, Lauren C. (2021), '*Brown Lake* by Samantha Lang', *Prototype*, 28 April, https://youaretheprototype.art/brown-lake/. Accessed 6 May 2021.

Hennessey, Kate (2021), 'Sprouting new life', *The Saturday Paper*. 5 June, p. 27.

Juhasz, Alexandra and Lebow, Alisa (2018), 'Beyond story: An online, community-based manifesto', *World Records*, 2, Fall, https://vols.worldrecordsjournal.org/02/03. Accessed 26 March 2021.

Juhasz, Alexandra and Lebow, Alisa (eds) (2021), *World Records*, special issue: 'Beyond story', 5, https://vols.worldrecordsjournal.org/05/01?index=1. Accessed 27 Jan 2022.

Kuzmanovic, Maja, Gaffney, Nic, Broglio, Ron and Nocek, Adam (eds) (2019), 'An introduction', *Dust & Shadow Reader #2: Attunement*, FoAM, p. 3.

Lang, Samantha (2021), *Brown Lake*, Australia: Prototype, https://youaretheprototype.art/brown-lake/. Accessed 6 May 2021.

MacDonald, Scott (2013), 'The Ecocinema experience', in S. Rust, S. Monani, and S. Cubitt (eds), *Ecocinema Theory and Practice*, AFI Film Readers, London: Routledge, pp. 17-41.

Mason, John (2002), *Researching your Own Practice: The Discipline of Noticing*, London: RoutledgeFalmer.

MECO Network (2019), *100 Atmospheres: Studies in Scale and Wonder*, London: Open Humanities Press.

Miles, Adrian (2014), 'Click, think, link: Interval and affective narrative', in M. Soar and M. Gagnon (eds), *Database | Narrative | Archive: Seven Interactive Essays on Digital Nonlinear Storytelling*, http://dnaanthology.com/anvc/dna/click-think-link-interval-and-affective-narrative. Accessed 13 May 2020.

Miles, Adrian (2016), 'I'm sorry I don't have a story: An essay involving interactive documentary, Bristol and hypertext', *VIEW Journal of European Television History and Culture*, 5:10, 31 December, https://www.viewjournal.eu/articles/ https://doi.org/10.18146/2213-0969.2016.jethc113/. Accessed 31 March 2021.

Murphie, Andrew (2014), 'Making sense: The transformation of documentary by digital and networked media', *Studies in Documentary Film*, 8:3, pp. 188–204.

Panse, Silke (2013a), 'Land as protagonist: An interview with James Benning', in A. Pick and G. Narraway (eds), *Screening Nature: Cinema Beyond the Human*, New York: Berghahn Books, pp. 60–70.

Panse, Silke (2013b), '*Ten Skies*, *13 Lakes*, 15 pools - Structure, immanence and eco-aesthetics in *The Swimmer* and James Benning's land films', in A. Pick and G. Narraway (eds), *Screening Nature: Cinema Beyond the Human*, New York: Berghahn Books, pp. 37–59.

Smith, Sable E., Story, Brett, Wilkerson, Travis and Juhasz, Alexandra (2021), 'Character limit', *World Records*, special issue: 'Beyond Story', 5, pp. 113–120.

Soar, Matt (2014), 'Making (with) the Korsakow system: Database documentaries as articulation and assembly', in K. Nash, C. Summerhayes, and C. Hight (eds), *New Documentary Ecologies: Emerging Platforms, Practices and Discourses*, London: Palgrave Macmillan, pp. 154–173.

Tsing, Anna L. (2015), *The Mushroom at the End of the World: On the Possibility of Life in Capitalist Ruins*, Princeton, NJ: Princeton University Press.

Tsing, Anna L., Swanson, Helen, Gan, Elaine, and Bubandt, Nils (eds) (2017), 'Introduction: Haunted landscapes of the Anthropocene', *Arts of Living on a Damaged Planet: Ghosts of the Anthropocene*, Minneapolis, MN: University of Minnesota Press, pp. G1–G16.

PART III

Emerging Technological Practices

CHAPTER 11

An Ecoaesthetic of Vegetal Surfaces: On *Seed, Image, Ground* as Soft Montage

Jussi Parikka and Abelardo Gil-Fournier

Introduction

This essay builds on our recent *Seed, Image, Ground* video project (2020) and how it entangles with our larger research project on surfaces. Both address the surface of plants, agriculture and the biosphere in relation to the surfaces of media, such as screens and images. The video essay was commissioned by the Fotomuseum Winterthur in 2020 and it is openly available at the institution's Youtube channel.[1] While the video is a central

[1] The url to access to it is https://www.youtube.com/watch?v=iMgijwDyq88 [Accessed on March 24, 2021].

J. Parikka (✉)
Department of Digital Design and Information Studies, Aarhus University, Aarhus, Denmark
e-mail: parikka@cc.au.dk

University of Southampton, Southampton, UK

A. Gil-Fournier
FAMU, Prague, Czech Republic

D. Binns and R. Najdowski (eds.), *Confronting the Climate Crisis*, Palgrave Studies in Media and Environmental Communication,
https://doi.org/10.1007/978-3-031-89606-4_11

reference point for this visual essay, our aim is not so much to theorize our own moving images and their juxtapositions and rhythms. Instead, in this article, we present a series of surfaces and scales that appear in and through the images. Images build upon images and this constitutes the practice-led approach to the temporal unfolding in the video. The video works as a temporal articulation of image surfaces across and upon living surfaces. In other words, we articulate the moving images as practices of surfaces, plants, and images: hence the central motif of the video essay and this accompanying text is to ask "what do images of growth look like?" How do such images operate in two-channel video practice that builds on Harun Farocki's work? Here we do not refer only to operational images (Farocki 2004), but also to the idea of a "soft montage" (Farocki 2009) which becomes a methodological reference point as one aspect of practice of images about and across images that can also be conceptualized as we do in this essay. The video and its relation to text are ecoaesthetic (Cubitt 2017) in the sense of dealing with the entanglement of media and environmental materials where the double articulation of images and green plant surfaces are the engine of our argument.

In this way, the video is a methodological exercise to articulate a problem space (Lury 2021)—a space where problems, methods, and concepts are dynamically rearticulated, and where the space of their composition is not a stable container where a thing is addressed but part of their making: "a problem is not given but emerges with-in and out-with a myriad sequence of actions or methods that (trans)forms the problem space." (Lury 2021: 6). This composition and this montage points to the ways methods are enacted only in the material practices that constitute the problem space as it unfolds as a living issue and tissue. Can this sequence Lury refers to be also a sequence of images? Could the actions be those of images? As the following visualization (Fig. 11.1) of the evolution of the timeline of our video essay show, the video's timeline as a problem space presents its fabrication as a surface. The cuts, the movements of the sequences, and other editing operations manifested in the figure, call for a dynamic understanding of surface as surfacing, built on multiple genealogies of instrumentation, technology, and cultivation.

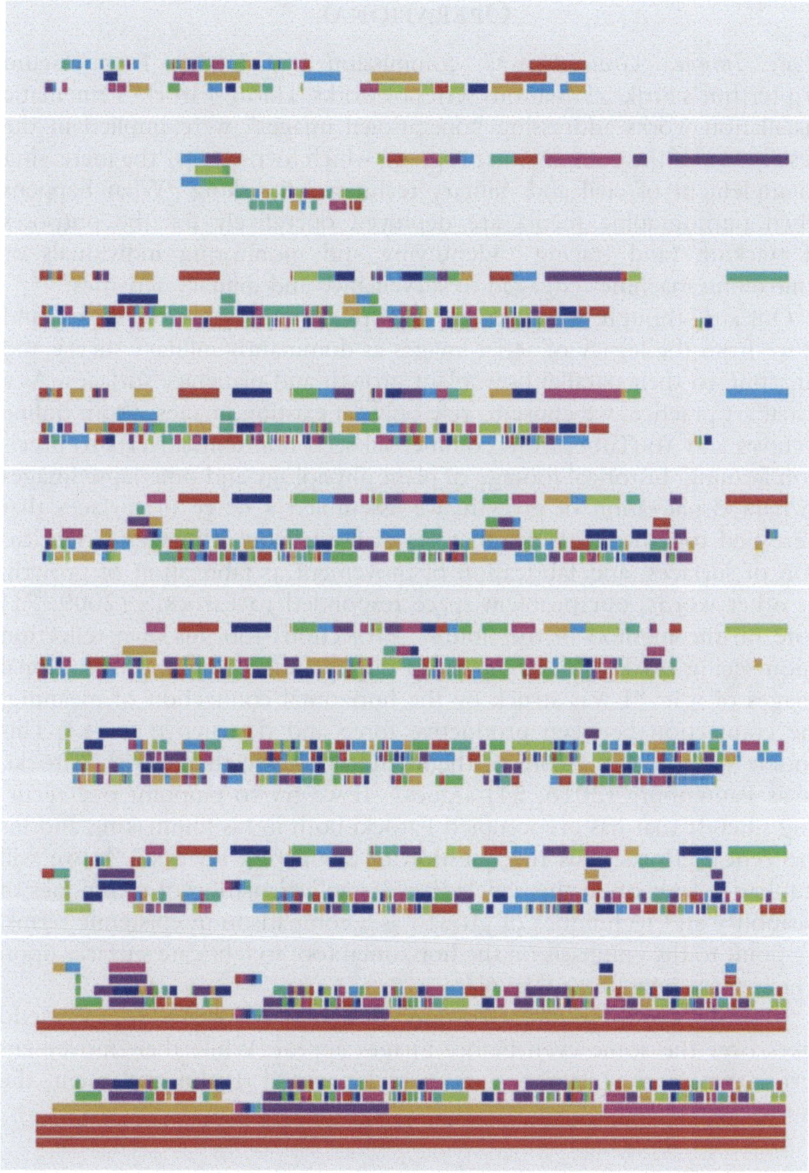

Fig. 11.1 Timeline as a problem space: visualization of the different versions of the *Seed, Image, Ground's* timeline, 2020. © Abelardo Gil-Fournier and Jussi Parikka

Operational

Seed, Image, Ground was commissioned for the Fotomuseum Winterthur's Strike/Situations series of works. Harun Farocki's cinematic installation works addressing "operational images" were implied in the description of the commission program, which focused on "the increasing entanglement of civil and military technology", asking "What happens when photographic media are deployed operatively for the purposes of tracking (and tracing), identifying and monitoring individuals or controlling machines engaged in surveillance and military activities?".[2]

Our aim, though, was to divert from the strictest sense of "strike" and move from the legacy of aerial images as dominantly military survey and targeting to their parallel uses: plant growth and planetary surfaces. As a visual art practice, we chose to rely only on existing images. From online archives and YouTube, from commercial seed bombing adverts to precision farming, historical footage of plant physiology and time-lapse images to data visualization of growth, we assembled a range of surfaces that were tied together with the rhetoric of fabrication of images as fabrication of surfaces, and fabrication of movement as fabrication of growth. In other words, our problem space responded to Farocki's (2009: 71) note on the method of the double (projection) too, his own reflection upon seeing his Eye/Machine series on machine vision and operational images of war: "I was struck by the horizontal connection of meaning, the connection between productive force and destructive force." This double bind is also a core aesthetic and epistemic theme with Farocki, or as Pantenburg (2016: 54) argues, "If we try to pinpoint one recurring interest that has preoccupied Farocki both in his filmmaking and his theoretical efforts, it is the question of comparing images." As we will emphasize later on in this article, besides an articulation of techniques of weaponry and techniques of growth as a comparison in epistemic terms, we point to the emphasis on the horizontal too: to fabricate surfaces upon which those operations take place.

In *Seed, Image, Ground*, the sound design by María Andueza Olmedo takes over the scene even before images appear. While their movement across the screen split screen space is the central rhythmic element, the

[2] See the webpage "Strike" at https://www.fotomuseum.ch/en/cluster/24/ (last accessed 6 April, 2021).

sound has its own narrative and non-narrative qualities. The remediation of the mechanical airlift of the helicopter rotors sets the backdrop of the cinematic 1970s. The whirling trope familiar from *Apocalypse Now* (1979) is replaced into this other 1970 cinematic maneuver of Andrei Tarkovsky's *Mirror* (1975) where the mystical wind that blows across the crop fields is created by the helicopter, carefully placed behind the camera. This sound speaks to a periodization of media archaeological combination of cinema and war (Virilio 1989) which is in the pace of the opening moments, though, quickly replaced with the whirling rotation of a seedling. This, then, provides a disjunction between operations of war and growth. Hence, the opening question of the voice over "how do you fabricate wind" is not about a poetic metaphysics, but one of specific situations of cultural techniques (Siegert 2015) of environmental management, to put it into administrative language. More than a question it is an expression of a problem, or even more accurately with Lury: it composes a problem in the methodological arrangement of images across and on surfaces (Fig. 11.2).

The video picks up on seed bombing as a motif that relates to specific techniques of enhancement of plant growth and reforesting. Seed bombing ranges in scale, most often it is discussed as a DIY technique for seeding derelict or hard to reach land and ground. But it is also something that features in histories, plans, and some actual projects of reforestation. The military history of the idea emerges from former RAF pilot Jack Walters' proposals and already 20 years ago it was developed as commercial idea by Aerial Forestation Inc. premising its large-scale viability on "Equipment installed in the huge C-130 transport aircraft used by the military for laying carpets of landmines across combat zones has been adapted to deposit the trees in remote areas including parts of Scotland." (Brown 1999). What's interesting is the relation between aerial power—and military logistics—as it is related to environmental techniques; the superiority of the aerial view as it has been outlined and the complexity of techniques that go into enabling this view.

As Saint-Amour (2013) argues in his take on aerial imaging during the First World War and the subsequent years, a complex set of arrangements and ways of seeing as ways of knowing—a bundle of interrelated operations—enabled the "planimetric view's construction from a diachronic series of images and moments." In short, one does not simply view from above, but it involved in those military operations a complex composition of scales, speeds, viewpoints and other forms of "applied modernism"

Fig. 11.2 Abelardo Gil-Fournier and Jussi Parikka, *Seed, Image, Ground*, 2020, Video stills. © Abelardo Gil-Fournier and Jussi Parikka

(Saint-Amour 2013) that worked through the figure of the photomosaic surface. The multiple shots taken were temporally and dynamically out of sync because of the different speeds of aeroplanes doing the imaging; hence the different scales needed to be operationally synchronized. Such techniques of making surfaces had also an impact on post-war territorial management. Indeed, the beauty of their applicability was to offer the illusion of simultaneity of areas of a territory stitched together into one surface where military techniques fed into different forms of aerially driven bureaucracy and design, such as urban planning. Fabricating surfaces became thus one key technique of knowledge established through aesthetic crafts, photography, and technologies of ascent: "A photographic form that revealed camouflage had itself been camouflaged, its contingencies disguised or levelled so that it resembled an idealised schematic as seen by a disembodied observer. Canonised by intelligence personnel, surveyors, police, entrepreneurs and city planners as the utmost in photographic realism, the mosaic was considered 'accurate' only in proportion as it bent the rules of optics away from perception toward orthogeny." (Saint-Amour 2013, 133) (Figs. 11.3 and 11.4).

In other words, cognizant of the trajectories of aerial view but also techniques of the surface, a misuse of military equipment (Kittler 2013) applies also to environmental management. We shift gradually from the operational sphere of the military to a constellation of images that speak of images:

(1) Plant growth modeled as an image;
(2) Growth/transformation expressed in moving images
(3) Territorial management as one of imaging: growth and management of territories as images and data come together.

We develop such points in our recent book Living Surfaces (Gil-Fournier and Parikka 2024) that addresses the interplay between vegetal surfaces and screen surfaces. That work dips in more detail into some historical scenes of nineteenth-century plant physiology and the use of images and apparatuses in laboratory practices. Besides working with science studies inspired methods with historical material, we extend the question of the plant surface into different scales: the definition of what is growth, and how images are incorporated into an understanding of growth—photographic images and diagrams, and the also historical

Fig. 11.3 Abelardo Gil-Fournier and Jussi Parikka, *Seed, Image, Ground*, 2020, Left side of video still. © Abelardo Gil-Fournier and Jussi Parikka

Fig. 11.4 Abelardo Gil-Fournier and Jussi Parikka, *Seed, Image, Ground*, 2020, Right side of video still. © Abelardo Gil-Fournier and Jussi Parikka

and more recent techniques of "time-lapse" with cinematic techniques entering the scene around fin-de-siecle; and how in more recent terms this is taken further with techniques of large-scale datasets but also making territories into data manageable images such as in precision farming as just mentioned.

The operational image has come to refer to images that do not primarily represent but are part of an operation (Farocki 2004); industrial and scientific images, images that function in systems, images that are for navigation and guidance or measurement, for example. Machine vision and AI is where the term has recently mushroomed. We, however, extend the use of the term, in order to: firstly, to engage with the operationalization of processes of light as they measure plants and become part of the epistemic operation (epistemic with very close goals related to industry and agriculture) and secondly, operationalizing also large-scale areas like farming areas as part of data-driven operations, reminding of a crucial theme often ignored in media and other research: infrastructures of urbanism extend to rural areas too including AI, blockchain, and other techniques dealing with datasets consisting of biological organisms and ecological areas such as in plantations and farming (see Wang 2020).

In our visual essay, we also shift gradually from the operational sphere of the military to a constellation of images that speak of images: in Farocki's terms a soft montage as a method of speculative surfacing. This surfacing is both a visual theme but it pertains to what we consider a central element of the articulating links that are being created, also then shifting the term soft montage to the question of temporal problem space (s).

A Soft Montage

Our images are found images; the video essay does not include any sequence filmed by ourselves. They are sourced from multiple uses: from commercial material, promotional, casual footage and less casual projects. But the project is not an archival project—that would stretch the term archive too far. It is a montage project, a project of problematics and composition. Removed from their original arrangement, yet not erasing all of their contexts, for us, the images are compiled and arranged into a split screen space where, echoing Blümlinger on Farocki's multiscreen projections, "the organisation of images appears much more important than the method used to record or generate them." (2008: 276).

As per our earlier note about the ecoaesthetic, the split screen is for the double articulation of image and landscape, image and territory. "Multiple images and visual contexts created independently of one another collide here," in Blümlinger's (2008: 277) words again. She continues: "the recorded images intersect and reflect one another." (ibid.) Technical images in plant physiology and agricultural management (including soil maps and other specialist techniques of early data visualization) constitute one reference point for the epistemological function of what images do in scientific practices. Such images are, we argue, not only ways to mark and measure and thus understand the enhanced ways of cultivation of growth, but they are also operational images as they prescribe a particular role still to images.

Split screen is however always more than two as it has its own force of synthesis as well as rhythm. History of cinema and video is one continuing experiment on this from Dziga Vertov's fragmented visual space of superimposition to many later examples of multiscreen projection formats (Beck and Bishop 2020). One can even speak of videographic cinema (Rosenkrantz 2020). Farocki (2009: 74) also refers to this lineage and such a practice of assembly through his notion of soft montage: "A montage must hold together with invisible forces the things that would otherwise become muddled." For Farocki, this thought continues in relation to his works on technologies of war, implying that montage is a method of articulating media archaeological continuities and discontinuities: "Is war technology still the forerunner of civil technology, such as radar, ultra short-wave, computer, stereo sound, jet planes? And if so, must there be further wars so that advances in technology continue, or would the simulated wars produced in laboratories suffice?" (Ibid.)

What does Farocki mean by "the things that would other wise become muddled"? Are the images there to clarify things, to set them straight? Or are they there to set linear things into their right, complex order: a problem that is composed as it unfolds surface to surface. The screens in a soft montage or the synchronized channels in a video installation provide cuts across historical periods and as such, have the quality of images composing and recomposing other images. Or, phrased differently: historical contexts infusing other historical periods; in some ways, soft montage is the work of (media) archaeology or genealogy (Parikka 2012; Elsaesser 2017), but across a set of electronic images and video (Fig. 11.5).

Fig. 11.5 Abelardo Gil-Fournier and Jussi Parikka, *Seed, Image, Ground*, 2020, Video clips displayed as dataset. © Abelardo Gil-Fournier and Jussi Parikka

Farocki (2009: 70) continues on *the double*, clarifying also our approach to soft montage as method about the synthesis of surfaces as surfacing: "There is succession as well as simultaneity in a double projection, the relationship of an image to the one that follows as well as the one beside it; a relationship to the preceding as well as to the concurrent one. Imagine three double bonds jumping back and forth between the six carbon atoms of a benzene ring; I envisage the same ambiguity in the relationship of an element in an image track to the one succeeding or accompanying it." Arranged images can function to open a field of problematics. Here, the split is also an addition: the one and the two, and the third and the fourth; not just on screen but also across multiple layers of surfaces of materiality that become known as they appear: both as objects of knowledge (laboratory practices) and as objects of intervention (agricultural surface). Thus, in our research, agricultural and laboratory surfaces entwine in the surface of the image, as wars and simulation of wars in Farocki's work. Beyond simulation, soft montage stands for their mutual synthesis: it is about their relational constitution or, in Bruno's (2014: 17) words, about "the substance of material relations" as they manifest on the surface of different media.

Farocki does not offer a clear definition of soft montage. Clarifying the stakes of the soft montage, Alter (2015: 151) proposes that it is more about a "general relatedness of images, rather than a strict equation of opposition produced by a linear montage of sharp cuts." Alter emphasizes that the soft montage differs from the more famous Eisenstein version and is defined by "a logic of difference." Alter (2015: 152) continues: "In this regard the cut of soft montage is synonymous with the conjunction "and," as multiple images are folded onto one another within the same spatial field, creating new configurations."[3] Alter's definition of soft montage points out that it can be read in proximity to an emphasis found in Gilles Deleuze's philosophy: the conjunctive *and* of the assemblage concerns the multiplication of difference, which still creates a field of resonances and thus it also creates a field of problematics; we can identify shots and reverse shots, but they also become speculative as they bypass narrative forms of historical time.

[3] This relates to Helga Lutz's research project in Germany that identifies elements of the double and the comparative as a media technique. "Media Dispositifs of Comparison: Harun Farocki's comparative practices in the context of media history" project description online at uni-bielefeld.de/(en)/sfb1288/projekte/c05.html. See also Pantenburg (2016).

Our work is also an assembly in this particular sense that nods to the historical archives as well as contemporary images. There is no simple causality that the images would express in what force impacts which thing, what surface becomes operative and which surface becomes data. That would be too crude of an assumption. Instead, the scenes relate to a *series* of images that form and formulate surfaces. When we are interested in the question of the plant and territorial surface, we engage through the image surface. When we are dealing with image (media) surface, it leads to ask how are they modelled, which material base and reference is apt when considering the contemporary ecoaesthetic situation. The images, thus, investigate surface inscriptions (Fig. 11.6).

Vegetal Filmmaking

The aerial view is likely to be a central part in the repertoire of techniques of air, but it also includes the idea of a large-scale territorial planning, measurement, and survey. It is through making the territory into an image that a particular operationalization also for growth becomes possible. As images that define or focus on space, we juxtapose different sets, contexts and archival sources: laboratory images of growth; a set of scientific procedures and images that are instrumental in such operations of measurement of plants since nineteenth-century plant physiology; seed bombing and regreening surfaces of growth; and precision agriculture that visualizes the surface anew based on processed data relations, even AI techniques, representing what we coin "platform ruralism": a mode of management of large-scale growth surfaces through an agglomeration and processing of data on corporate and other platforms that then feed back onto the surface as agricultural operations whether mechanical or chemical. Platform urbanism (Barns 2019) has come to refer to such data platforms of circulation of gathering, circulation, and analysis of data in smart cities, but this extends it to non-human living contexts of plants and growth: a variety of data-points that do not refer to the usual sphere of media in urban consumer use, but to its long back trail in rural lands and environmental conditions.

Writing about the use of multiple projections in the turn of the twenty-first-century video art, Liz Kotz argues that "images find their natural home in the emerging world of 'information architecture', in which every building functions as a kind of image and every built surface promises to contain some type of networked display" (2008 383). While the argument

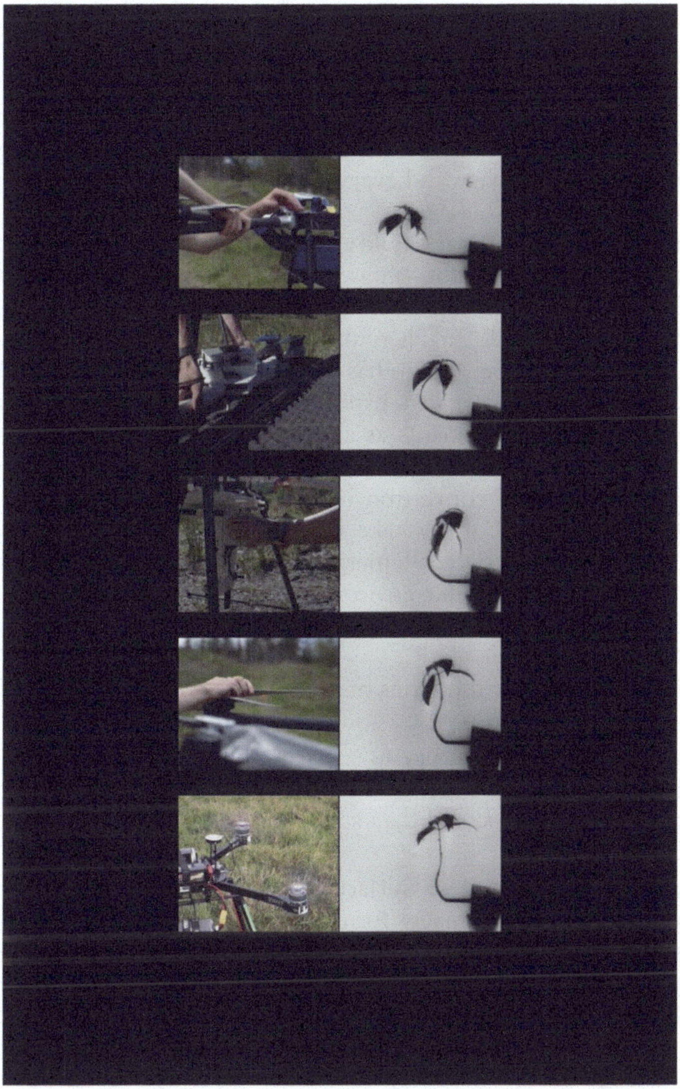

Fig. 11.6 Abelardo Gil-Fournier and Jussi Parikka, *Seed, Image, Ground*, 2020, Video stills. © Abelardo Gil-Fournier and Jussi Parikka

typically assumes the surfaces of the architectural building or the urban space, as in Giuliana Bruno's landmark work (2014), in *Seed, Image, Ground* we extend this particular interweaving to other built environments such as those in aerial reforestation, landscape restoration, and agriculture. Building on Kotz (2008), here every vegetal surface would function as a kind of image and every landscape would promise to behave as a networked screen.

"'The future of the movies,' said Abel Gance, 'is a sun in each image'" (Bozak 2011:34). While the phrase by Gance recalls what Bozak has described as the "elemental solar origins" of cinema and its fossil fuel materiality (Bozak 2011: 35), here it opens up into a space where images and vegetal surfaces meet. Uhlin has written about "vegetal filmmaking," "where vegetal being is given expressive form through a filmmaker's particular deployment of plants as a structuring element of their work" (2015: 202) while taking into account at the same time that both images and plants carry out a conversion of solar energy (2015: 203). In this context, and beyond Gance's take on solarity, the surface of our split screen as well as the related "emerging world of information architecture" that pervades the built environments of platform ruralism, makes images and plants coalesce as surfaces accelerated by the same diagram (Fig. 11.6).

Hence, we articulate the series of surfaces as series of image problems:

Territorial surfaces, and plant surfaces
Plant surfaces, and agricultural surfaces
Surveys, landscapes, and data surfaces

As problem spaces, these surface spaces compose each other. The sequences and actions of images form a topology of the problem at hand: what is a surface, how is seed bombing a surface operation, and how do images feature in this rethinking of the territory both from above and as data. Lury's (2021) methodological argument then becomes one reference point for our work while the composition of images and the composition of discursive concepts are part of this work of soft montage.

Funding Details The research for this article was supported by the Operational Images and Visual Culture project (2019–2023), funded by the Czech Science Foundation, 19-26865X.

References

Alter, Nora. 2015. "Two or Three Things I Know about Harun Farocki" *October* 151, Winter 2015: 151–158.

Barns, Sarah. 2019. *Platform Urbanism: Negotiating Platform Ecosystems in Connected Cities*. Basingstoke: Palgrave.

Beck, John and Bishop, Ryan. 2020. *Technocrats of the Imagination: Art, Technology, and the Military-Industrial Avant-Garde*. Durham: Duke University Press.

Blümlinger, Christa. 2008. "The Art of the Possible: Notes About Some Installations by Harun Farocki" In *Art and the Moving Image: A Critical Reader*, 273–282. London , New York: Tate Publishing.

Bozak, Nadia. 2011. *The Cinematic Footprint: Lights, Camera, Natural Resources*. New Brunswick, N.J.: Rutgers University Press.

Bruno, Giuliana. 2014. *Surface: Matters of Aesthetics, Materiality, and Media*. Chicago: The University of Chicago Press.

Cubitt, Sean. 2017. *Finite Media. Environmental Implications of Digital Technologies*. Durham: Duke University Press.

Elsaesser, Thomas. 2017. "Simulation and the Labour of Invisibility: Harun Farocki's Life Manuals" Vol 12 (3): 214–229.

Farocki, Harun. 2004. "Phantom Images" *Public* 29: 12-24.

Farocki, Harun. 2009. "Cross Influence/Soft Montage" In *Harun Farocki. Against What? Against Whom?*, edited by Antje Ehmann and Kodwo Eshun, 69-74. London: Raven Row & Koenig Books.

Gil-Fournier, Abelardo and Jussi, Parikka. 2024. *Living Surfaces: Images, Plants, and Environments of Media*. Cambridge, MA: The MIT Press.

Kittler, Friedrich. 2013. "Rock Music: A Misuse of Military Equipment" In *The Truth of the Technological World. Essays on the Genealogy of Presence*. Translated by Erik Butler, 152–164. Redwood, CA: Stanford University Press.

Kotz, Liz. 2008. "Video Projection: The Space Between Screens in Art and the Moving Image" In *Art and the Moving Image: A Critical Reader*, 371–385. London , New York: Tate Publishing.

Lury, Celia. 2021. *Problem Spaces. How and Why Methodology Matters*. Cambridge: Polity.

Pantenburg, Volker. 2016. "Working Images: Harun Farocki and the Operational Image" In *Image Operations : Visual Media and Political Conflict*. Edited by Jens Eder and Charlotte Klonk, 49–62. Manchester: Manchester University Press.

Parikka, Jussi. 2012. *What is Media Archaeology?* Cambridge: Polity.

Rosenkrantz, Jonathan. 2020. *Videographic Cinema*. London: Bloomsbury.

Saint Amour, Paul K. 2013. "Photomosaics. Mapping the Front, Mapping the City" In *From Above: War, Violence and Verticality*. Edited by Peter Adey, Mark Whitehead and Alison J. Williams, 119–142. London: Hurst.

Siegert, Bernhard. 2015. *Cultural Techniques. Grids, Filters, Doors, and Other Articulations of the Real*. Translated and edited by Geoffrey Winthrop-Young. New York: Fordham University Press.
Uhlin, Graig. 2015. "Plant-Thinking with Film: Reed, Branch, Flower" In *The Green Thread: Dialogues with the Vegetal World*. Edited by Patrícia Vieira, Monica Gagliano, and John Ryan, 201–217. Lanham: Lexington Books.
Virilio, Paul. 1989. *War and Cinema*. Translated and edited by Patrick Camiller. London: Verso.
Wang, Xiaowei. 2020. *Blockchain Chicken Farm*. New York: FSG Originals/ Logic.

CHAPTER 12

Speculative Visions: Machine Learning, Photography and the Climate Crisis

Rebecca Najdowski and Daniel Palmer

What might imagery generated from machine learning teach us about the Anthropocene? In the context of the climate emergency, social media feeds and news broadcasts are filled with imagery of a warming planet like red smoke-filled skies, bleached coral, and receding glaciers. These visual signs serve as a warning of dangers to come, of imminent possible scenarios and extinctions, yet necessarily arrive late—as traces of disasters that have already taken place. In this chapter, we consider a different temporality for picturing environmental collapse beyond that of documentation. Specifically, we explore how image-based machine learning—built on large datasets of photographs—can offer new ways to envisage the climate emergency. Analysing a range of future-oriented artistic projects and approaches (including Refik Anadol's *Quantum Memories*,

R. Najdowski (✉) · D. Palmer
School of Art, RMIT University, Melbourne, Australia
e-mail: rebecca.najdowski@rmit.edu.au

D. Palmer
e-mail: daniel.palmer@rmit.edu.au

D. Binns and R. Najdowski (eds.), *Confronting the Climate Crisis*, Palgrave Studies in Media and Environmental Communication,
https://doi.org/10.1007/978-3-031-89606-4_12

Anna Ridler's *The Shell Record*, Rebecca Najdowski's *Deep Learning the Climate Emergency* and Tega Brain, Julian Oliver and Bengt Sjölén's *Asunder*), our argument is that machine learning can both disorient and expand our field of vision in the face of the climate crisis, potentially amplifying our willingness to act. Given the politics of climate change is fought around data modelling, we argue that a similarly speculative imagery is now required.

Introduction

Photography has long been used to record the impact and damage that humans have had on the Earth's environment. In the context of the climate emergency, social media feeds and news broadcasts are filled with imagery of a warming planet—red smoke-filled skies, flooded streets, parched landscapes, bleached coral and receding glaciers. For obvious pictorial reasons, photographs of extreme weather events like floods and droughts have featured heavily (Palmer 2013). The subject of environmental collapse provides such dramatic imagery as stranded polar bears, burnt koalas, melting glaciers and flooded cities, a documentary imagery of effects that has inevitably become clichéd. But photography has also been the key medium for environmental criticism due to what media theorist Ana Peraica describes as 'its capacities for documenting time and reporting across large distances, thus, its ability to register slow processes and distant changes' (2020: 100). Take 'before-and-after' photography (Palmer 2013), a common and arguably quite effective visual method used by documentary makers and environmental campaigners (including popular figures like David Attenborough). A photograph of a glacier from fifty years ago can be shown next to an image of the same scene today to show how much it has receded. Peraica rightly points out that such documentation may even be incidental; in the era she calls the 'Photographocene', photographs 'document the continuous human impact of the environment, intentionally or not' (2020: 107).

Photographs of climate change thus relate to temporality in interesting ways. On the one hand, photographs necessarily arrive late, as traces of disasters that have already taken place. As Julie Doyle puts it, 'photographs of glaciers represent temporally the *already seen* effects of climate change' (2007: 129, original emphasis). And yet, as before-and-after as well as time-lapse photography illustrate,[1] through drawing attention to slow-moving changes, photographs can serve as a warning of dangers to come, of imminent possible scenarios and extinctions (Richardson 2020). In this sense, as Peraica (2020: 106) argues, photographs of melting glaciers can be conceived as 'anticipatory images'.

In this chapter, we seek to reconsider the documentation of climate change by exploring new forms of speculative imagery opened up by algorithmic photography and computer visualisation. The online visualisation tool This Climate Does Not Exist (2021) by researchers at Mila, the Quebec Artificial Intelligence Institute offers a simple example. A website (https://thisclimatedoesnotexist.com) invites the user to type in an address and choose an environmental impact. A photographic image of the precise location from Google Street View is then served up, while machine learning is utilised to render a realistic filter of climate change induced flood (from torrential rain or rising sea levels), engulfed in fog (from pollution) or red smoke haze (from fires). Prompting people to type in their own home address, the website asks: 'Can you imagine these kinds of disasters happening in your backyard?' It goes on:

> For the planet to tackle the climate crisis, we all need to act as though our homes were directly affected. We have harnessed AI to build a tool that lets you imagine the impact of the climate crisis we face, one address at a time. (This Climate Does Not Exist 2021: n.pag.)

Describing itself as an 'AI-driven experience based on empathy', *This Climate Does Not Exist* seeks to simultaneously raise awareness about possible scenarios as well as promote the achievements of AI to visualise the future ('These images are generated by AI. But the environmental disasters they portray are real') (This Climate Does Not Exist 2021: n.pag.). Building on the before-and-after genre, an existing scene is augmented with an uncanny future catastrophe.

In this chapter, we explore four artists and collectives whose projects speak to the future in relation to the climate emergency. Since first drafting this paper, AI-generated imagery has advanced at a frenetic pace. Text-to-image generators such as DALL-E, Midjourney and Stable Diffusion, capable of generating photorealistic yet synthetic images from prompts, have become available for general use and positioned AI imagery in the mainstream. However, in this chapter we are specifically interested in artist's use of photographic datasets to offer a different temporality for picturing environmental collapse beyond the conventions of documentation. Given that the global heating happening right now is invisible and its future effects can only be anticipated via predictive statistical modelling, we propose that speculative imagery provides a necessary disorientation and expansion of our fields of vision in the face of the climate crisis.

Machine-Based Learning, Temporality and Artistic Engagement

The subset of artificial intelligence (AI) known as machine learning has quickly become important to many artists. In contrast to rule-based AI which entails directly programming computers to be 'smart', machine learning approaches computational intelligence as something that can be learned (Audry 2021). Machine learning uses networked algorithms that take inspiration from the networks of the human brain with the idea of mimicking biological neural activity. Comprised of three components—data, the training process and the model—at its core, machine learning uncovers patterns. In this chapter, the projects that we discuss primarily use General Adversarial Networks (GANs), a form of deep learning. GANs first appeared in the machine learning field in 2014 with work by computer scientist Ian Goodfellow and researchers at his lab, and since then artists have explored the aesthetic and conceptual potential using these networks. Conveniently, the internet has offered a ready source of images for mass data extraction, 'akin to a natural resource' (Crawford 2021: 106). As Sofian Audry points out:

> [deep learning's] emergence since the beginning of the millennium is inseparable from the increased access to raw computing power—in particular, due to the development of graphical processing units (GPUs) incidentally pushed by the game and cinema industries—and the exponential growth of data, thanks to the explosive expansion of the internet as a platform for mass social media. (Audry 2021: 4)

In the most basic sense, GANs use a sample dataset, learn the patterns in that set, and then approximate what exists as new outputs. In the training process, two competing networks basically play a game with each other: the generator creates an image attempting to fool the discriminator into identifying it as a real image from the training dataset, and the discriminator attempts to determine if the image is real or fake. This iterative process pushes the generator to produce more and more accurate images as the training continues. Through the training, the machine generates what is called 'latent space', a mathematical representation of compressed data where each point in multi-dimensional space represents an individual image. The training process for GANs is not reliant on labels in the way discriminative models used for classifying are, yet the labelling of images is still crucial for dataset generation. Many of the datasets used in image-based machine learning training rely on how images are labelled or tagged to create a collection of images of a given subject. The training often requires thousands of images, and this is typically done by 'scraping' internet sources for images using keywords. An unlabelled image will remain invisible in the collecting process.

The visual results of machine learning are often discussed in terms of the uncanny (Zylinska 2020: 81; Hertzman 2020). The computer generates abstractions that may be more or less legible to human actors, raising a broader paradox about algorithms and images observed by Louise Amoore:

> There are two curiously twinned accounts of contemporary algorithms in relation to regimes of sight and vision. The first is that algorithms operate on a plane in excess of human visibility and at scales that are inscrutable to the human. The second is that algorithms themselves have an enhanced capacity to visualize the invisible, to see, scan, and search volumes and varieties of data heretofore unavailable to human senses. (Amoore 2020: 15)

It is this second quality that is often celebrated by artists working with AI, as in the case of Refik Anadol below. But as artist and geographer Trevor Paglen (2016) points out, images generated through machine learning are primarily designed for other machines. They belong to 'AI systems [that] have appropriated human visual culture and transformed it into a massive, flexible training set' (Paglen 2016: n.pag.). Scientists, and artists, can 'evaluate' the training by creating outputs of data stored within the latent space of the models. But our ability to 'read' these images is limited. Paglen emphases that understanding the largely invisible machine-machine visual culture requires unlearning the way that humans see: 'we need to learn how to see a parallel universe composed of activations, keypoints, eigenfaces, feature transforms, classifiers, training sets, and the like' (2016: n.pag.).

One can question if the images produced by machine learning are photographic, even as they have the appearance of photographs, given no 'light writing' is involved. Competing perspectives on the condition of photography in the context of machine learning highlight a nebulousness in 'photographic' discourse. For example, on the one hand, Andrew Dewdney (2021), in his book *Forget Photography*, argues that photography has been supplanted in the shift from optical to algorithmic images, favouring the term 'the networked image'. Conversely, Paglen proposes an expanded definition of photography that *includes* machine vision devices, image-based data, and the 'seeing-practices in which they are enmeshed' (Paglen 2014: n.pag.). Notably, Dewdney highlights an ontological-epistemological paradox: 'at the moment of photography's technical replacement by the screen, algorithm and data flow, photographic cultures proliferate like never before' (2021: 3) where 'the image still communicates, speaks and performs to humans and it continues to do so through a cultural code of representation' (2021: 162). Computational images generated through machine learning models, trained on photographic datasets, carry with them a 'representational logic' (Rubinstein and Sluis 2013; Dewdney 2021) of photography. While we acknowledge the paradoxical condition of these images, we see them as part of an ongoing 'desire' in and of photography (Batchen 1997: 2016).

As Amoore has argued, 'when algorithms learn by inductively generating outputs that are contingent on their input data, they are engaging experimentally with the world' (2020: 12). However, numerous scholars have argued that machine learning can be dangerously deterministic; the system cannot acquire knowledge beyond the data that it is fed for training, making it a tool which reinforces what it already knows, using correlation and pattern recognition (Impett 2017; Crawford 2021). As the photographic theorist Katrina Sluis suggests, citing Impett, a 'positive feedback loop or algorithmic "way of seeing" is reinforced by search engines, photography apps and camera software' that 'inform dominant visual hegemonies whilst simultaneously learning from them, and therefore fortifying and reinforcing them' (Sluis 2019: n.pag.). For this reason, it is particularly important, as Audry suggests, that artists are in a position to 'suggest alternative ways of engaging with machine learning systems and imagining our relationship with them now and in the future' (2021: 16). This requires new skills. For instance, artists can directly intervene by tweaking datasets, models and the evaluation function of AI software, in what amounts to 'a form of collaboration between the artist and the machine' (Audry 2021: 16).

In this chapter, we are interested in understanding how experiments by contemporary artists with image-based machine learning might offer new ways to envisage the climate emergency, beyond the conventions of documentation. Our analysis of various examples by artists working with machine learning will pay particular attention to different forms of temporality that result. We do not suggest that the calculative architecture of the algorithms behind machine learning produce one particular temporal mode. Datasets have their own temporalities; as 'the residual contingencies of durational time, with all the past lived moments supplying norms and anomalies for the algorithm to learn' (Amoore 2020: 17). However, machine learning is clearly more future-oriented than the 'has been' tense of conventional photographic practice (Barthes 1981: 77). In this sense, and in their contingency and fallibility, machine learning provides for models of creative speculation that offer a compelling parallel to the most important way that global heating is understood and communicated today—namely through climate modelling.

Refik Anadol's Digital Memories of Nature

Turkish-born, Los Angeles based Refik Anadol is among the best-known artists working with AI today. His prolific studio now includes more than a dozen designers, architects and coders and unashamedly operates like a commercial design studio, in collaboration with major technology companies. Anadol's work has involved advanced experiments with AI using supercomputing power, applied to cultural archives and real-time data such as that related to transportation, wind or sea surface data (*Bosphorous*, 2018). We are particularly interested here in his work that involves large photographic datasets of nature. These range from a collaboration with Samsung, *Artificial Landscapes* (2021), which promote the idea of 'data as a pigment' and where 'visual memories of nature' resemble an impressive screensaver, to more conceptually ambitious artworks such as *Quantum memories* (2020), commissioned by the National Gallery of Victoria (NGV) in Melbourne for their Triennial exhibition and now part of their permanent art collection. A spectacular 'data sculpture', displayed on an enormous 10 × 10 metre media wall positioned in the monumental forecourt of the fortress-like gallery, this work is based on an aggregation of photographs of the natural world. Although pioneering in its use of quantum computing, the work includes all the swirling, colourful, dynamic and abstract qualities that have come to characterise Anadol's work—including an impressive *trompe l'oeil* effect in which flat video screens appear to present a three-dimensional image spilling over their edges. The work 'draws upon a dataset of more than two hundred million nature-related images from the internet' (NGV Triennial 2020: n.pag.), processed in real time. Immersive and spectacular, visitors to the gallery were transfixed in awe, reaching for their iPhones and the language of the sublime to describe the overwhelming volume and scale of the data visualisation, accompanied by a hypnotising soundtrack also composed in collaboration with a generative algorithm (Fig. 12.1).

Anadol (2019) refers to his mode of data visualisation as 'data dramatization', evoking its capacity to provoke an emotional response.He speaks about his work as 'a poetic exploration of what is hidden in the mind of machines'—by which he alludes to their potential for consciousness, or even 'dreaming', rather than a critical understanding of biased algorithms or datasets (Anadol 2019). Anadol is hardly the first to ask such questions; notions of hallucination and dreaming are frequently discussed in relation to the aesthetics of AI, given the seemingly real perception of something that does not actually exist.[2] Specifically, *Quantum Memories* proposes a machine in the near future with a form of consciousness and capacity to imagine an alternative *natural* universe based on what it has learned from humanity. Thus, according to gallery didactics, *Quantum Memories* turns 'the visual data that flows around us every day into an artwork that represents our collective memory of nature' and 'can be considered both an alternate dimension of the natural world and a radical visualisation of our digitised memories of nature' (NGV Triennial 2020: n.pag.).

Fig. 12.1 NGV Triennial (2020) installation view of Refik Anadol (designer); Refik Anadol Studio, Los Angeles (design studio), *Quantum Memories*, 2020. Commissioned by the National Gallery of Victoria, Melbourne

Here, Anadol builds on a conventional association between photography and memory and a curiously naïve notion that 'the machine doesn't forget' to ask: 'How can we create a library of the future, to remember better' (Anadol and LeAmon 2020). But the notion of 'our' digitised memories is questionable, given photographs are not memories and the online archives his team have scraped are inevitably skewed to the (over) developed world. Anadol suggests it offers a 'new narrative for humanity' (Anadol 2021). But neither this redemptive future vision nor 'our collective memory of nature' extends to considering human impacts or the climate emergency. One newspaper reported that Australian landscapes made up part of the dataset (Te Koha and Carmona 2020). But since viewers have no access to these source images, what is collectively forgotten is that they either depict visibly polluted scenes or 'picturesque denials of human impact[s] on the environment' (Peraica 2020: 107).

More recently, Anadol has produced a series of nature-based works under the banner *Machine Hallucinations* which he contextualises in terms of environmental collapse. For example, Anadol posted the following Instagram caption during the Californian fires of late summer 2021:

> Dear friends, sharing our latest AI Cinema experiment work for you in this heavy wild fires days. *Machine Hallucinations – Nature Dreams* (2020) using 68,986,479 photographic landscape memories of our Earth with machine intelligence. Hope these are not our last memories! (@refikanadol 2021: n.pag.)

Such melancholic allusions to the end of 'photographic landscape memories of our Earth' may not in fact contradict Anadol's celebration of post-human machine visions of nature. Likewise, *Machine Hallucinations—Coral* (2021), generates patterns based on 1,742,772 images of coral and aims to bring to the impact of climate change on coral die-offs. The work appeals to a notion of photographic evidence, but the absence of dates and locations associated with the photographs in the dataset leads to an intentionally impressionistic result. Remarkably, in discussions about Anadol's work, despite his use of hundreds of millions of photographs—many mined from image-centric social media 'to understand how others feel' (Anadol 2019)—photography itself is never addressed. We are told that *Quantum Memories* 'draws upon a dataset of more than two hundred million nature-related images from the internet, which are processed using quantum computing software' (NGV Triennial 2020: n.pag.). But the neutral language conceals hidden decisions within the selection of data and how the algorithms are applied. Like nature, photography is here mere data for visual dramatisation, as part of a highly Instagrammable artwork that in turn generates more photography for machines.

Anna Ridler's Hand-Crafted Datasets

By contrast to Anadol, the British artist Anna Ridler creates 'self-generated' datasets, to develop very different narratives about nature and the climate emergency. For example, *Myriad (Tulips)* (2018), is an installation of 10,000 hand-labelled photographs of tulips that were later used as the dataset for *Mosaic Virus* (2018 and 2019). As Ridler notes on her website:

> By choosing to make the dataset an artwork it draws attention to the skill, labour and time that goes into constructing it, whilst also helping to expose the human element in machine learning, usually hidden by algorithmic processes. (Anna Ridler Website, n.d.)

The categorisation of datasets is a process that fixes the flux and flow of images, which Ridler (2019) describes as 'a spatio-temporal segmentation of the world'. Just as humans make meaning of our surroundings, in machine learning the classifying process relies on human labour to label and curate. Because of the immense number of images required, computer scientists outsourced the labour of categorising images to low-paid workers on Amazon's Mechanical Turk site to develop the foundational datasets. The most widely known example is ImageNet, a dataset originally published by Li Fei-Fei and her collaborators in 2009, which used tens of thousands of Mechanical Turk workers for the labelling process. Ridler's self-generated datasets, crafted from her own photographs, serve to counter this evisceration of labour (Fig. 12.2).

In the same way that individual photographs can feel 'dated', so too can datasets. They are a record of a specific time, and once created can be used again and again for different research and creative purposes. Yet the datasets remain suspended in time, like silver halide particles on the surface of an analogue photographic print. While the training, model and the subsequent visual output are contingent—images and videos can be produced from the models' learning at any point—the dataset functions

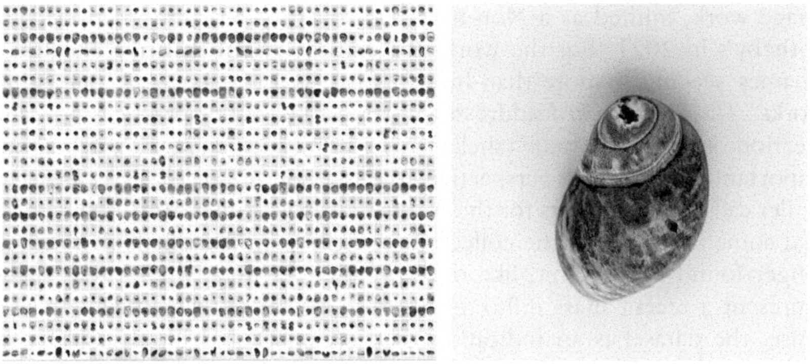

Fig. 12.2 Anna Ridler, *The Shell Record*, 2021, PNG image of dataset and GAN generated video still

like a kind of archive. This archival quality of the dataset is taken up in Ridler's collaborative art project with designer Caroline Sinders, *Cypress Trees: A Beginning* (2021, in progress). The work depicts morphing and shifting cypress trees along with what seems to be audio of waves lapping at land. The images that formed the basis of this generative artwork are photographs taken by Sinders in the swampland of Louisiana and Mississippi. There, cypress trees are increasingly impacted by the influx of saltwater in their habitat caused by dredging for oil and gas where the accumulating salinity begins to kill the trees. It is this precarity of the protected and iconic cypress trees that contributed to the artists considering their work as a type of archive; a historical record of trees that in the near future might not survive the impact of environmental disruption (Sinders 2021). In the generative artwork, we do not encounter the archive directly; instead, what we see is the model's 'interpretation' of the photographic dataset—a vision of possible but unreal cypress trees at the edge of uncertainty.

However, in many of Ridler's machine learning projects, photographic datasets are presented alongside generated artefacts from the models. While many other machine learning results are untethered from their training source, in a way severed from their referent, Ridler is explicit in connecting the dataset as a system of knowledge to the algorithmically produced images. For example, *The Shell Record* (2021) is a PNG file of a grid of photographs of shells (the dataset) and a GAN-generated moving image work, minted as a Non-fungible Token (NFT) and auctioned by Sotheby's in 2021. For the work, Ridler collected shells from the River Thames seeking a more-than-human history. Like much of her other works, *The Shell Record* addresses historical collecting practices and the creation of economic value (shells having often served as currency). More importantly, from our perspective, the species of shells serve as what Ridler calls 'time-markers for the Anthropocene' (Ridler 2021). She notes that some of the shells she collected are from marine animals that are no longer found in the river, like once prevalent oysters, while other shells represent a recent mass influx of invasive species (Ridler 2021). In this sense, the dataset is an indication of the climate and biodiversity crisis and points to a temporal scale that is both geologic and human. Unlike Anadol's abstract 'dreaming', Ridler's algorithmically generated images thus stem from and incorporate indexical datasets whose real locations and histories are crucial to their meaning.

Rebecca Najdowski's Visualisation of the Climate Emergency

A number of other artists are using machine learning and photography to scrutinise how nature and technology are enmeshed. For example, US-Australian artist Rebecca Najdowski's *Deep Learning the Climate Emergency* (2022) uses pervasive photographs depicting the effects of a heating planet to generate moving image artwork. The impetus of the project came about as the artist noticed the phenomenon—which she participated in—of social media posts featuring eerie suns, altered and addled by smoke from megafires in southeast Australia and California in 2019 and 2020. In order to explore this emerging aesthetic of the climate crisis, she created a dataset from these images with the idea of exploring how artificial intelligence 'saw' this phenomenon. From images of strange smoke affected suns, the project expanded to include common tropes of the climate emergency such as wildfire, bleached coral, drought, and melting glaciers. Najdowski sought to understand if such familiar images of the climate emergency, ones that have perhaps lost their potency through repetition, could be reconfigured through machine learning to produce something new.

In a machine learning context, the training datasets for *Deep Learning the Climate Emergency* are relatively small at around 1,000 images for each subject. As a dataset, the theme of suns through smoky skies is reminiscent of artist Penelope Umbrico's meditation on authorship and the collective facet of photographic culture. For her well-known *Suns from Sunsets from Flickr* (2006–ongoing), Umbrico aggregates images of sunsets posted to the social media site and presents them as a massive grid of hundreds of prints of the cropped orbs. Like much of her work, it is a project that speaks to the practice of ubiquitous image sharing and, in a sense, prefigures the creation of image datasets for machine learning art.

Visual artworks that use machine learning are part of a genealogy of imaging technology that has enabled perception beyond the human scale—what Walter Benjamin called 'another nature' in *Little History of Photography* (Benjamin [1931] 1999: 510). Like photographic and film technologies, images generated from the layered algorithms of GANs allow for a new perception that was previously unavailable to humans. Matteo Pasquinelli and Vladan Joler describe the 'generative modality' of machine learning—its predictive capacity to create patterns from the statistical models—as something that can 'reveal what the model has learnt, i.e. to show how the model "sees the world"' (2021: 1274). In the case of *Deep Learning the Climate Emergency*, the moving images result from an exploration of latent space, sometimes called a 'latent walk', where the output reflects a traversal between selected points in the multi-dimensional space, each representing an AI-generated image. A latent walk allows us to picture computational learning and, in a way, peer into the models' 'perception', or patterning, of the world (Fig. 12.3).

In *Deep Learning the Climate Emergency*, the animation between vectors in latent space results in a morphing of suns, cracked earth,

Fig. 12.3 Rebecca Najdowski, *Deep Learning the Climate Emergency*, 2022, stills from GAN generated video

calving glaciers, scenes of wildfire, and bleached coral. Some of the visual renderings of the models are representationally accurate to our world, while others take on an uncanny quality, producing what Aaron Hertzmann describes as the 'visual indeterminacy' of some GAN-produced imagery, where the aesthetic 'arrangements and textures are not discrete but continuous' (2020: 425). This visual indeterminacy and constant (re)generation relates to the strange and uncanny condition of living within the temporality of climate change, where the consequences of historical activities haunt the present and near future. Refracted through the prism of machine learning, Najdowski's artwork is a rendering of the shapes and textures of the Anthropocene.

Tega Brain, Julian Oliver and Bengt Sjölén's Critique of Automation

Our final example, *Asunder* (2019), a collaborative artwork by New York based Tega Brain, Berlin based Julian Oliver, and Stockholm based Bengt Sjölén, addresses the application of AI to climate change modelling in a direct and critical way. Commissioned and originally exhibited in *UNCANNY VALUES: Artificial Intelligence and You* as part of the Vienna Biennale for Change, the work is a three-channel video projection that incorporates satellite imagery and Community Earth System Model (CESM) to simulate the Earth's past, present and future climate states. *Asunder* mounts a critical response to the emerging application of AI to critical environmental challenges, by adopting it as 'a literal proposition' (Asunder Website 2019). Based on available data and modelling, the artists have created a fictional 'environment manager' in the form of a supercomputer that gathers satellite, climate, topography, geology, biodiversity, population and social media data in real time. Under the guise of mitigating global heating, the supercomputer manager then simulates future changes to real sites, including the Artic, a Brazilian rainforest, Silicon Valley and Vienna, with often preposterous outcomes. With consideration of human impacts demoted in favour of a computational planetary logic, whole cities are relocated, nations combined, coastlines straightened and rivers moved. Like Anadol's work, *Asunder* speculates about a future based on computer dreaming. However, the self-conscious absurdity of the results serve to warn of the blind application of AI to the challenges of the climate emergency and the assumption that such operations are politically neutral.

Asunder is far from a simple parody of AI. Instead, it seeks to make visible what are otherwise invisible models, and by positioning 'both landscape and ecosystem as computational surface' implicitly questions the ideological framing of the environment as a system (Brain 2018). Fusing machine learning image techniques with traditional climate modelling—including historical images of specific sites alongside simulations of what those places may look like in the future—statistical correlation is placed alongside observation and theoretical modelling. As Brain observes, a key limitation in climate modelling is that we only have a couple of hundred years of data to work with, and the models fail to account for how environments can change in unexpected ways. From this perspective, machine vision and technologies of remote sensing are not in themselves invalid, but deeper questions need to asked about how they are currently used, and can be used differently 'to amplify environmental encounters and the arts of noticing our ecological entanglements' (Brain 2018: 160) (Fig. 12.4).

Asunder is a complex example of the operative and speculative use of photographic imagery, employing photographic data from satellite imagery to question what handing over decision-making to computational processes might look like. It builds on an earlier work of Brain's, *Deep Swamp* (2018), which also subjected real environments to the control of a software agent. That system worked by capturing and analysing photographs of physical wetland environments created in a gallery in real time and compared them to different computer vision models, each trained on thousands of images of wetlands scraped from social media

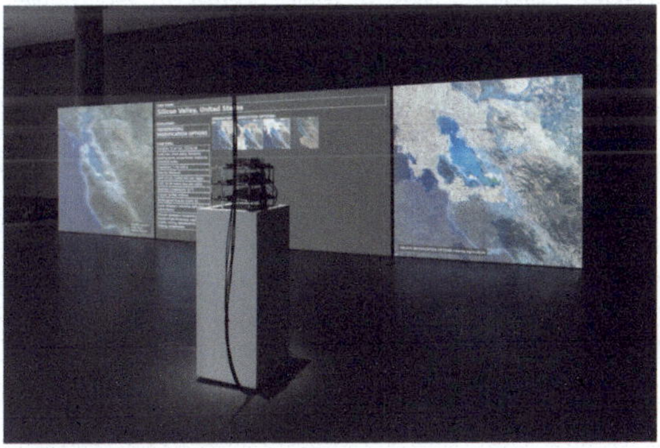

Fig. 12.4 Tega Brain, Julian Oliver & Bengt Sjölén, Asunder, 2019, Three channel video-projection, satellite imagery, CESM climate model, multi-processor computer and custom software, at Transmediale 2020

platforms such as Flickr. Based on this data, the software agent was able to change the growing conditions (create mist, change light levels and so on). Like *Asunder*, the work employed the evidential logic of photography within a complex chain of processes to invite a public conversation about the uses of such automating technologies and methods. Both works recognise that the climate crisis involves a disruption of patterns and ask how this connects with machine learning and its function of pattern extraction, recognition and generation.

Learning from Disorientation

The 'slow violence' of environmental catastrophes (Nixon 2011: 2) and the implementation of AI at all levels of life are perhaps the two most transformative developments currently impacting life on this planet. Moreover, a number of thinkers have pointed to their relationship; the mega computing power required for AI and machine learning necessitates the intensive use of natural resources and energy. As Roel Dobbe and Meredith Whittaker highlight, AI and the climate crisis are detrimentally entangled, with the exponentially growing carbon intensive technology field lacking transparent accounting for its carbon footprint (Dobbe and Wittaker 2019: n.pg.). Artists approaching the climate emergency with AI are therefore compelled to confront this paradox. Nevertheless, as Joanna Zylinska argues, 'AI-driven art that responds responsibly [could] help us expand the current imaginary, allowing for the emergence of new visions and vistas for the world to come' (2020: 45). As Malcolm Miles once suggested, art must go beyond the imagery familiar to us from mass media, 'not by being more realistic or more beautiful—both outdated criteria—but by changing the conditions in which nature is produced' (2010: 24).

Timothy Morton (2018) argues in his meditations on 'the strangeness of living in an age of mass extinction' that all art, all imagery, produced today can in some sense be viewed in the context of environmental collapse. However, the art projects we have explored in this chapter specifically respond to Peraica's claim that 'The time of our documentation of life extinction, including our own slow dying into the image has somehow arrived, as the number of images produced exceeds reality and obscures our relationship to it' (2020: 108). Both the scale of image-data and the scale of our climate predicament is of a magnitude beyond human comprehension. Like global heating, machine learning and the photographic-and-algorithmic enabled images that it generates could be considered a 'hyperobject', an assemblage so 'massively distributed in time and space relative to humans' that we cannot fully grasp it (Morton 2013: 1). The artists discussed in this chapter use the spatiotemporal expanse as a condition that shapes their projects, whether in celebration, ambivalence or critique. Through harnessing machine learning for artistic purposes, these projects open up relations with machine-generated and operative images. As Paglen (2016) suggested was necessary in the age of operative images, these artistic approaches begin to enable a new kind of seeing that humans must learn in order to understand machine-machine image culture.

The images produced by neural networks can be conceived as a post-human form of sensing, even a preparatory course to adapt humans to a post-human world. Despite their often mimetic quality, these images are of an elsewhere. They are somehow of this planet, and yet not entirely. While we can consider how these images might be speculative in the sense that they depict possible but not real scenes, they are equally intriguing as a visualisation of the blackbox of the algorithmic models. Pasquinelli and Joler suggest that machine learning can be considered 'an instrument of knowledge magnification that helps to perceive features, patterns, and correlations through vast spaces of data beyond human reach' (2021: 1263). As a predictive process, machine learning necessarily relies on the past—built into the training datasets is the 'that-has-been' (Barthes 1981: 77), a feature deeply embedded within the strata of conventional photographic temporality. However, machine learning and its non-human visualisations can amplify and alter images of the climate emergency, (re)generating the photographic archive for new audiences.

In this chapter, we have analysed four projects that each explore the limits and potential of machine learning and photographic imagery in different ways. We have argued that while each have something to offer, the more critically oriented projects are both speculative and demystifying. Beyond the sublime abstractions of Anadol's work, Ridler focuses our attention on the individual elements of the datasets while Najdowski draws attention to indeterminacy as a defining feature of machine learning. Meanwhile, Brain, Oliver and Sjölén use photographic imagery as part of larger assemblage of data elements to question the planetary scale logic of AI as it intersects with climate modelling. Given the politics of climate change are formed by global data modelling, we argue that such experiments in speculative imagery help to educate as well as productively unsettle our fields of vision. Beyond simply drawing attention to the latest climate-influenced disasters, or revealing changes to natural environments over time, photographic archives are here put in the service of the future in a way that privileges the perspective of the non-human world, and underlines humanity's inseparable entanglement with 'nature'. Such non-human speculative visions also build up critical literacy around machine learning and its possible place within photographic projects. And they seem especially pertinent in light of Peraica's ominous warning that the time of the Photographocene 'will end when none are left to monitor this cumulative process' (2020: 107).

Notes

1. See, for example, James Balog's *Extreme Ice Survey*, an ambitious multi-year time-lapse project of receding glaciers initiated in 2007 (www.extremeicesurvey.org).
2. See, for example, Google's DeepDream (www.deepdreamgenerator.com) and Memo Atkin's *Learning to See* (2017) (www.memo.tv/works/learning-to-see).

References

Amoore, Louise (2020), *Cloud Ethics: Algorithms and the Attributes of Ourselves and Others*, Durham: Duke University Press.

Anadol, Refik (2019), 'In the Mind's Eye | Refik Anadol', YouTube, 13 June, https://youtu.be/TdI1zER_K18. Accessed 11 February 2022.

Anadol, Refik and LeAmon, Simone (2020), 'Triennial conversations | Quantum memories', YouTube, 20 December, https://youtu.be/Sp4vXxbLMHU. Accessed 11 February 2022.

Anadol, Refik (2021), 'NGV Triennial | Refik Anadol', YouTube, 19 April, https://youtu.be/oFsjVtmnbS0. Accessed 14 February 2022.

Anadol, Refik (@refikanadol) (2021), Instagram, 8 August, https://www.instagram.com/p/CSRrxmyIRTv. Accessed 11 February 2022.

Anna Ridler Website (n.d.), 'Myriad Tulips', http://annaridler.com/myriad-tulips. Accessed 15 February 2022.

Asunder Website (2019), 'About', https://asunder.earth/#description. Accessed 16 February 2022.

Audry, Sofian (2021), *Art in the Age of Machine Learning*, Cambridge, Massachusetts: The MIT Press.

Barthes, Roland (1981), *Camera Lucida: Reflections on Photography*, trans. Richard Howard, New York: Hill & Wang.

Batchen, Geoffrey (1997), *Burning With Desire: The Conception of Photography*, Cambridge, Massachusetts: The MIT Press.

Benjamin, Walter ([1931] 1999), 'Little History of Photography' in M. Jennings, H. Eiland and G. Smith (eds), *Walter Benjamin: Selected Writings 1931–1934*, vol. 2, translated by Rodney Livingstone et al., Cambridge, Massachusetts: Belknap, pp. 507–530.

Brain, Tegan. (2018), 'The Environment is not a System'. *APRJA*, 7:1, pp. 152–165.

Crawford, Kate (2021), *Atlas of AI: Power, Politics, and the Planetary Costs of Artificial Intelligence*, New Haven: Yale University Press.

Dewdney, Andrew (2021), *Forget Photography*, London: Goldsmiths Press.

Dobbe, Roel and Whittaker, Meredith (2019), 'AI and Climate Change: How they're connected, and what we can do about it', *AI Now Institute*, 18 October, https://medium.com/@AINowInstitute/ai-and-climate-change-how-theyre-connected-and-what-we-can-do-about-it-6aa8d0f5b32c. Accessed 15 December 2021.

Doyle, Julie (2007), 'Picturing the Clima(c)tic: Greenpeace and the Representational Politics of Climate Change Communication', *Science as Culture*, 16:2, pp. 129–150.

Hertzmann, Aaron (2020), 'Visual Indeterminacy in GAN Art', *Leonardo*, 53:4, pp. 424–428, https://doi.org/10.1162/leona01930.

Impett, Leonardo (2017), "Ways of Machine Seeing', 14 June, CRASSH, https://www.crassh.cam.ac.uk/blog/ways-of-machine-seeing/. Accessed 2 March 2022.

Miles, Malcom (2010), 'Representing Nature: Art and Climate Change', *Cultural Geographies*, 17:1, pp. 19–35.

Morton, Timothy (2013), *Hyperobjects: Philosophy and Ecology after the End of the World*, Minneapolis and London: University of Minnesota Press.

Morton, Timothy (2018), *Being Ecological*, Cambridge, Massachusetts: The MIT Press.

NGV Triennial (2020), 'Triennial Profile | Refik Anadol', National Gallery of Victoria, https://www.ngv.vic.gov.au/multimedia/triennial-refik-anadol/. Accessed 15 December 2021.

Paglen, Trevor (2014), 'Seeing Machines', Fotomuseum Winterthur, 13 March, https://www.fotomuseum.ch/en/2014/03/13/seeing-machines. Accessed 28 February 2022.

Paglen, Trevor (2016), 'Invisible Images (Your Pictures Are Looking at You)', *The New Inquiry*, 8 December, https://thenewinquiry.com/invisible-images-your-pictures-are-looking_ at-you. Accessed 28 February 2022.

Palmer, Daniel (2013), 'Photography, Technology, and Ecological Criticism: Beyond the Sublime Image of Disaster' in L. Lester and B. Hutchens (eds) *Environmental Conflict and the Media*, New York: Peter Lang, pp. 75–90.

Pasquinelli, Matteo and Joler, Vladan (2021), 'The Nooscope manifested: AI as instrument of knowledge extractivism', *AI & Society*, 36:4, pp. 1263–1280, https://doi.org/10.1007/s00146-020-01097-6.

Peraica, Ana (2020), 'Photographocene: The past, present and future in the photography of the environment', *Philosophy of Photography*, 11:1&2, pp. 99–111, https://doi.org/10.1386/pop_00030_1.

Richardson, Michael (2020), 'Witnessing the Anthropocene: affect and the problem of scale', *Parallax*, 26:3, pp. 339–359, https://doi.org/10.1080/13534645.2021.1883298.

Ridler, Anna (2019), 'Sorting Things Out: Dataset and Classification', YouTube, 12 November, https://youtu.be/wQmqZdTSES0. Accessed 27 March 2022.

Ridler, Anna (2021), 'The Shell Record | Natively Digital: A Curated NFT Sale | 2021', Sotheby's, n.d., https://www.sothebys.com/en/buy/auction/2021/natively-digital-a-curated-nft-sale-2/the-shell-record. Accessed 24 April 2022.

Rubinstein, Daniel and Sluis, Katrina (2013), 'Algorithmic Photography and the Crisis of Representation', in M. Lister (ed.), *The Photographic Image in Digital Culture*, New York: Routledge, pp. 22–40.

Sinders, Caroline (2021), 'Inside Futurelab: Cypress Trees, A Beginning – guest presentation by Anna Ridler & Caroline Sinders', YouTube, 21 April, https://youtu.be/offIGn8qQbo. Accessed 15 January 2022.

Sluis, Katrina (2019), 'Photography Must Be Curated! Part Four: Survival of the Fittest Image', *Still Searching: An Online Platform for Photographic Discourse*, Fotomuseum Winterthur, 20 December, https://www.fotomuseum.ch/en/2020/01/23/survival-of-the-fittest-image. Accessed 15 December 2021.

Te Koha, Nui and Carmona, Bianca (2020), 'How Refik Anadol created giant LED screen artwork Quantum Memories for NGV Triennial', *The Herald Sun*, 18 December, https://www.heraldsun.com.au/entertainment/arts/how-refik-anadol-created-giant-led-screen-artwork-quantum-memories-for-ngv-triennial/news-story/c32626d401784c7f606dd96adc007ca9. Accessed 15 December 2021.

This Climate Does Not Exist (2021), 'Home page', https://thisclimatedoesnotexist.com/. Accessed 5 December 2021.

Zylinska, Joanna (2020), *AI Art: Machine Visions and Warped Dreams*, London: Open Humanities Press.

CHAPTER 13

Wasted Bodies Against Ruined Landscapes: How Mass Effect and The Last of Us Depict the Death of Humanity, but Not Neoliberalism

Stephanie Farnsworth

Empty landscapes, toppled buildings, and desolate towns are the images that dominate climate catastrophe fictional media, a genre gaining greater prominence and attention as the world begins to acknowledge the realities of the crisis that we are in. Yet, in these media delineations, often lacking is the visceral devastation: Where are the bodies? Where are the visual markers of death and decay? In media about climate catastrophes from *The Day After Tomorrow* (2004) to *Don't Look Up* (2021) there have been scant depictions of the body horror that the end of climate stability represents, with rare exceptions such as the use of forced blood donations ('blood bags') in the *Mad Max* franchise, and the litany of dead bodies

S. Farnsworth (✉)
University of Sunderland, Sunderland, UK
e-mail: Stephanie_Farnsworth@sunderland.ac.uk

D. Binns and R. Najdowski (eds.), *Confronting the Climate Crisis*, Palgrave Studies in Media and Environmental Communication,
https://doi.org/10.1007/978-3-031-89606-4_13

Rick runs into in season one of the zombie apocalypse show *The Walking Dead*. While Rick encounters piles of bodies, not left by zombies but by the futile attempts at control and order by the military, few narratives are this scrutinizing of the flat-line of death. Some video games have broken this trend, ensuring the player is forced to walk among wasted bodies, ruined by a climate crisis and sometimes apocalypse. There are two game series in particular that have dared to do this: *The Last of Us* (Part I 2010, Part II 2020) and the original *Mass Effect* trilogy (2002, 2010, 2012). *The Last of Us* depicts a post-apocalypse world where microbial infections have led to people transforming into mutants (specifically, a form of zombie) that desire to infect and murder all who cross their path. In *Mass Effect*, the main character Shepard discovers a galactic threat—a fleet of organic and synthetic ships that have wiped out countless civilizations before. Shepard however, is met with scorn as the galactic Council refuses to believe that such a dangerous threat can exist, only for the reapers to arrive and to harvest genetic material of all alien species, in an attempt to wipe out every civilization. Trillions of lives are lost in the course of the final *Mass Effect* game. Both game series are bound by the temporal experiences of apocalyptic doom, reflected in a destructive climate but resulting specifically in the wasting of bodies. This chapter will examine the wasting of bodies in fictional representations of climate catastrophes caused by neoliberalism.

Both our climate crisis and the apocalypses explored in both series are caused directly by the inherent philosophy of neoliberalism, a structure that promotes the element of competition as the key marker of a society's success and what Monbiot (2016) compares to a "biological law," almost as though we are genetically coded to pursue every ambition we have. The pursuit of our dreams at any cost has come at a high price for our environment when consumerism has led to resources being gobbled and a shift from unionization protecting workers since the industrial age, to a decline of agitation as workers enter a beleaguered, worn-down state of constant work in part due to our networked lives (Wellman 2006). This chapter will particularly draw upon the works of sociologist and architectural expert Sennett for the examination of the idea of creeping space, and the reclamation of the environment. Transhumanism (Haraway 2006; Alaimo 2020) is a theme and a modality that will be employed due to the disrupted bodies present in both series: bodies wasted beyond their potential, and consumed within different spheres.

Video games can be a forum for expressions and debates around ethics, values, or visions of political manifestos but so tethered to neoliberalism are we, that time and again worlds that need to be rebuilt after destruction are remade in the same foundations of what has already failed. Cycles of late-stage capitalism will thus be explored as a recurring theme within this text, as bodies and landscapes are wasted and left to ruin, in games where the player ultimately has little choice or agency over the nature of the environmental landscape and must find a way to navigate and exist among the ruins and decay.

A Late Capitalist Wasting of Bodies

The phrase "wasted bodies" refers to the idea of to two principles: Humans reduced to a premature death (which, in neoliberal terms is a tragedy for the early loss to the workforce), and the visceral signs of death and decay that can disrupt or empower capitalist and hegemonic function. The horror of the dead zombie-like "clickers" in *The Last of Us* is twofold: The repulsion of a premature death from an unknown and unstoppable disease, and the symbol of the decayed body encapsulating the breakdown of a capitalist society. They are visual markers of the precariousness of this late-stage capitalism. In this series, the breakdown of late-capitalism has not led to a utopian resistance, but the struggle to form any society at all in its wake. Pandemics and wars have a common scene that is tethered to their situations: Temporary, or makeshift facilities to treat the injured or otherwise afflicted. In the case of a virus, these mini-institutions arise either from the desire to separate a contagion from the rest of the population, or because healthcare services become so overrun that patients spill over into these temporary areas. Worldwide, this was evidenced recently due to the COVID-19 pandemic that saw temporary hospitals being built in China and the United Kingdom (among others) to deal with the wave of rapid infections at the start of 2020. However, in the case of war, and particularly of the war illustrated in *Mass Effect*, temporary healthcare facilities are thrown together due to battles moving and the injured needing urgent treatment, and often because local facilities have been destroyed by invading forces, leading to a need for immediate replacement services built up from any limited resources that are available. The positioning and construction of these services are fundamental to the concept of wasted bodies; this is the care that people receive before a potential and sometimes anticipated premature death. This is where the wastage finalizes.

The wastage of bodies (and life) inevitably falls under concepts of Foucault's biopolitics. However, while there is a general understanding that this relates to the shifting away from monarchist systems and capital punishment as discipline to systems of governance and bodily order, there are claims that biopower and biopolitics have shifting meanings in Foucault's writings making it difficult to engage in anything other than their broad concepts (Kristensen 2013: 11). But, at the conceptual level, is an understanding of a society that has shifted into the biopolitics of medical normalization and a biopolitics that has seen the management of populations including in relation to the economy and opportunities (Levina 2009). *The Last of Us* and *Mass Effect* examine the breakdown of population control, and depict the wasted landscapes and bodies that mass death results in.

Richard Sennett, a sociologist specializing in architecture, examined the significance of these facilities to public health and the management of bodies. During the 1832 June Rebellion in Paris, there was a widespread outbreak of cholera. The cholera-related death of a parliamentarian who opposed the monarchy was a contributing factor to the violence between different factions that summer, while the city was managing a severe public health crisis. The philosophical conflict over democracy and the attempt to deliver greater rights for citizens, plunged the city into a wider debate on the order and structural organization of France. This may be reflected in some of the public health decisions taken, as temporary hospitals were set up out of the remains of a former market and brothel. In this new facility, patients were given beds that were arranged in lines under a glass roof so that the sun beamed down upon them; this was believed to help the sick as there was still the idea of the divinity of the power of God and the Lord's healing capabilities, this faith translated into a literal belief that the sun could disinfect people (Sennett 2018: 22). Temporary hospitals then are not just areas that pop up to treat the wounded, but are facilities crafted to offer care at a holistic and architectural level. In the final game of *Mass Effect*, the player listens to medics giving orders to citizens on how to treat wounds over the radio. There are no medical facilities remaining, even temporary, as the overwhelming power of the reapers has seen their destruction.

The Last of Us also contains temporary medical facilities—predominantly made up of tents—scattered in what was once controlled refugee areas, but they have been abandoned years later as the entirety of the United States became overrun with infected clickers, bringing an end to the previously-enforced order of late-capitalism. These temporary facilities were never taken down, because the entire population of the United States was forced to flee the virus, and millions were left for dead or infected. These formerly safe healthcare zones are a reminder of the collapse of a system that simply couldn't cope with the demands of a decaying and wasting population. Occasionally, in these areas, the player will find a wasted body, or an empty space completely devoid of all life, giving a haunting quality due to the teeming life that would have once been present.

In *Video Games As Culture*, Muriel and Crawford (2018) argue that video gaming is a material experience that splits the player into several places: The physical space, and the virtual space. Muriel and Crawford state this is a specifically embodied experience (85). This experience when playing *The Last of Us* emphasizes the horror of the climate catastrophe, the player-avatar navigates through wasted spaces, trying not to disturb wasted bodies as zombies, while the player occupies the physical realm, the gaming industry itself driven by the entrenched consumerist ideas that are systematic to our society. The player is in a time of before the catastrophe *and* after.

In the time of climate catastrophe that the player now occupies in the material, offline world, the entrenched consumerism of gaming is apparent by the modality of play. It is almost impossible now to be a gamer without also being networked. Even games that are not live service, such as *The Last of Us's* main campaign, encourage players to have internet access so that patches may be released to improve the game post-launch. As growth in populations has seen cities build and spread, we have begun to occupy virtual worlds and time: these reflect growing tensions and anxiety around the inevitable bust that will take place without serious action to tackle climate change caused by rampant late-stage consumerist driven capitalism. We long to escape in virtual worlds, away from the grind and anxiety, and yet we continue to wade through post-apocalypse stories (Poole 2017: n.pag), warnings about what happens if we check out of the fight to protect climate in favor of chasing consumerism. The player, to a point, is wasting their own potential as we sit playing hundreds, even thousands, of hours of content; are we educating ourselves through stories and do they lead to action, or is this another expression of consumerism as we download the latest high-stakes role playing games?

Landscapes and Borders

Landscapes are one the most important components to climate catastrophe stories: we see the visceral ruin of capitalism and our society laid bare when skyscrapers are dilapidated, when domed churches have collapsed and when debris litters the road making a route impassable for the player. But the landscapes we see are inextricably linked to bodies, and raise the fundamental question of where do the bodies go, and what happened to them?

As Sennett focused much on the archeology of communities, essentially, it was therefore impossible for him, or this chapter to ignore the work of Heidegger whose work in hermeneutics and phenomenology saw him forever associated with what is known today as place and space (Meek 2022: n.pag). While Heidegger set out the fundamental ways in which place can be tied to the body, his theories were very much influenced by his politics: and therefore his close association with Nazism. As Sennett pointed out, in order for Heidegger to find a place to dwell and establish roots, he sought the Black Forest—a place where he could exclude others, particularly Jewish people (2018: 124).

The body is closely connected to the land in Heidegger's worldview, believing we are embodied in the environment which led to the focus on *dasein* ("dwelling"). The idea of deliberate and desired separation from others is deeply tied to a cultural phobia or discomfort of *the other*. For Heidegger, this manifested in antisemitism. Heidegger was affiliated with the Nazi Party and their propaganda was wrapped up in reclaiming the "fatherland" from foreign influences (a guise for the persecution of even German Jewish people). For racists and others with deep-rooted prejudices such as Heidegger, the countryside becomes a safe area to dwell with visitors not seen for miles, sometimes with the horizon empty except for fields and forest. A border is therefore established, where the dweller can enforce their own national identity and supremacist code within their fence posts. The collapsing of borders is a central tenant to (post) apocalyptical science fiction stories, and in particular those stories that incorporate zombies. The structures that we cling to and organize our lives through suddenly come under strain and often collapse, overwhelmed with the zombie threat. Capitalism quite simply stops functioning.

In apocalypse societies, a common thread is the fear of others. These narratives focus on the question: Without structures or order to enforce peace, who will protect us? Joel fears people and contact throughout *The Last of Us* as he leaves the community he knows to travel with Ellie, a young and immune girl determined to find a community force who are trying to cure the disease afflicting the population. The one place that represents true safety is a community called Jackson, with high walls, and people guarding the gate. When Joel and Ellie arrive there, they are threatened before it is revealed that Joel is the brother of one of the men who runs the community. As Sennett expanded: "There are two ways to shun alien others: Flee them or isolate them. Each way can take a built form" (2018: 126).

This was the principle behind Heidegger dwelling in the Black Forest, and Joel adheres to this philosophy of shunning all alien others and trying to isolate all threats that are possible. Sennett reflected upon the historic persecution of Jews, dating well before the rule of the Nazis over Germany, when in the 1500 s communities were placed within the walls of Spanish ghettos but were forced to stay there. This gave Jews what Sennett terms "placed-based rights" (2018: 132) in that they were protected from violence by those very walls. However, those rights were dependent on borders; borders that gave safety but not any security, that granted relief from persecution but not freedom from oppression or suppression. Borders give a false impression of safety, but they are a symbol of instability, of a desperate need to regain control and deploy a police force, for borders do not function if there is nobody there to monitor for breaches. Ironically, in *The Last of Us Part Two*, Ellie's girlfriend, Dina is Jewish and lives within the borders of the settlement of Jackson. She agrees to leave that safety but importantly, as a Jewish character, she is granted the freedom to choose. For the majority of the first game, Joel spends his time admiring or even seeking out such community spaces: this is because the landscape is littered with wasted bodies.

When Wasted Bodies Become Part of a Networked Machine

The Last of Us and *Mass Effect* are joined together by player-avatars wandering through various ruined landscapes, however in the latter game, the player has far more overarching power in trying to prevent future disaster by ensuring that various NPCs (and different species of characters) are subjects to Commander Shepard's orders. The avatar-body has vastly diverse levels of power and autonomy—by way of player agency, through wandering the landscapes and finding spaces to dwell (Heidegger, 2018: 124) to the experience as a subject of biopower—but they both operate under the remnant need for bodies to be managed under a form of capitalism and neoliberalism, even in the face of structural annihilation.

Throughout *The Last of Us*, the child Ellie wishes to be taken to the group known as "The Fireflies" so that her blood can be studied as she is the only known human to have been bitten by an infected being without "turning" or dying. However, as Joel acts as Ellie's guardian and grows attached to her, his goal separates from the child's. Joel longs to settle down in a safe community (which his brother helps to run) or else take their chances on their own. His desire stems from the understandable fear of the infected, but also from a fear that if anyone discovers Ellie's genetic importance, she may be in grave danger. It is then little surprise that when Ellie does find The Fireflies, they immediately sedate her and plan to harvest her genetic matter to be able to create a cure for the virus—a process that would kill Ellie. Humanity would be saved from this wasted world, but Ellie's life would be lost. The player has no say in the moral choice to save Ellie. Joel will always fight to protect her. The player's only autonomous choice is whether to kill all of the medics before leaving the hospital with an unconscious Ellie. The player therefore has to choose between wasting bodies; ending the potential of these medics to work in a pandemic because, as Joel perceives it, they may remain a threat to Ellie's claim to life. Player agency is limited, and what must be confronted by the player is the emotional challenge of the separation between the character we occupy and ourselves—a diegesis less individualized and potentially less likely to reveal "reflective eudaimonic emotional experiences" (Cole and Gillies 2019: n.pag). Material constraints over decisions "detracts from a sense of agency" (Mateas 2001: 145); we see Joel's complex thought process throughout the game but can only alter minor moments in the climax of the game. It has been argued that player agency naturally conflicts with game narrative and *The Last of Us* is at odds with the position of scholars such as Murray who infer that a central pleasure of digital environments is seeing the results of our decisions (Tanennaum and Tanenbaum 2010). Joel's actions can be a fraught moment for the player, who is denied the chance to save the world but instead is committed to wasting (at least some) bodies.

The zombie, like climate change, is described as a man-made disaster (Wasson and Alder, 2014, p.5), a symbol of late-stage neoliberal capitalism collapsing under the weight of its own lust for consumerism and ravaging its own systems and resources. Additionally, Comaroff and Comaroff (2013) postulate on the zombie as a representation of the fear of collapsing borders. They studied fear and tension around foreign nationals in South Africa and the belief that jobs were being stolen from the domestic workforce (2013: 139). Zombies represented the tension between what was portrayed in the media as an invading force putting nationals at risk, and the depiction of zombies as an all-consuming labor force that was single-minded in its task to take over every space (2013: 140). The zombie mutant then acts an agent of capitalism: In *Mass Effect* reaper "husks"—people turned into zombie-like creatures that serve the reapers—are set upon killing and destroying people who have not been indoctrinated into reaper control. In *The Last of Us,* clickers seek to bite and infect other beings so that they will begin a transformation into a zombie, a process that happened to Ellie's young friend Sam. Clickers and husks are linked by their need to contaminate, control, and infect other beings. The husks in particular, are set upon deliberately recruiting as many people as possible so the reapers can harvest genetic material. The reapers do not seek to immediately destroy, but to utilize husks as sleeper agents that gradually distort and infiltrate populations. The process for transforming into a husk is slow—it takes the course of several months for major characters Saren and The Illusive Man to be fully brought under control by the reapers and transform into husk-like beings. The manifestation of the dead as workers, is seen through their loss of life coupled with a lack of decay. The husks are people, transformed and changed into workers, who no longer represent living autonomy. But in the horror of this, the decay of humanity does not translate to biology and the husks have upgraded technology, and simply put, do not contain or produce bio-waste to evoke the true abject horror of death (Kristeva 1982: 3).

Michel Foucault put forward the notion of biopolitics as the need ""to ensure, sustain, and multiply life, to put this life in order" (1976: 138) and that the era of biopower was a shift from sovereign power, to centralized elected systems, and simply put—a shift from the divine right of kings deciding upon capital punishment for traitors to the crown, to an era of prison systems, careful employment management and ensuring the dominance of certain populations (1995: n.pag). Foucault's philosophy focused on analyzing the structural realm that dominates neoliberal societies, and looked at how the practices of death had changed. Death became managed, not something associated directly to the state, as lives were extended (and so the working age of people was extended, too), lives were carefully managed and then deaths were staged through funerals and wakes, as rituals of death. Life expectancy rose significantly in capitalist societies in the West (particularly the United States), and so early death was treated as a tragedy for the missed potential—a life that had fallen outside the realms of centralized conveyor belt control (Cooper 2008: 7). The zombies in *The Last of Us* and *Mass Effect* show the breakdown of neoliberalism through their deaths, by confronting and challenging the expected methods of body management. However, they bring a new system of a relentless and inescapable form of late-stage capitalism; a workforce where their bodies have no autonomy whatsoever, the only job is to ensure more workers join the cause and their incapacitated bodies become a desired state.

Throughout *The Last of Us* the player observes the visceral shock of bodies laying to waste; there are dead and decaying bodies throughout the game, but most shockingly of all are the bodies that are scattered inside buildings, so covered in the microbial infection that it is as though they are plants growing out of the wall. It can, at first glance, appear as though nature has reclaimed the buildings and reclaimed the bodies, after the world's population have decimated the world. Yet bodies fixed to former spaces of work and drudgery are decidedly unnatural, and what is highlighted here is the horror of the way neoliberal capitalism has destroyed bodies. Nature did not reclaim the world: the world simply collapsed beneath the weight of its own unsustainability.

The most gruesome depiction of wasted bodies is portrayed in *Mass Effect 3*, when Shepard is transported to the galactic capital, the Citadel space station, now under reaper control and being used as a processing plant. Victims are transported from Earth to the station, and murdered so their genetic material may be used to build more reapers. The player must walk by piles of corpses laid to waste. The transportations of entire populations to become part of the oppressive, relentless machine in pursuit of constant progress is a blunt metaphor for the horrors of climate catastrophes induced by late-stage capitalism. This relentless line of movable bodies is depicted in other media such as *Star Wars: Andor* (2022) which examines the prison-industrial complex or via entire homeless communities monitored through high-technology surveillance and moved out of Night City in *Cyberpunk 2077* (2020) by the police. It was biopunk (or biological dystopian) scholar Lars Schmeink who identified that in the relation to biocapital human lives must be managed (by regimes and authorities), and transformed into biological commodities which determine economic value (2016: 226).

Schmeink (2016) highlighted the key feature of the management of bodies: the material value that their labor can bring. This ties into Bocock's theories of productivity; under neoliberal capitalism, that works, essentially, in two mechanical ways: The more bodies that are applied to a task then the greater productivity, and for bodies to be able to work and feel happy with their lives (and therefore not challenge the system) they need to be able to feel like they can gain, or purchase, new items to bolster social status and security (1993: 3). This is the fundamental basis for a consumerist society. However, it creates a beleaguered body (Mitchell and Snyder 2015: 40), one in need of constant maintenance due to the extraordinary levels of work it is put through by the system. This is why we live under a structure where healthcare has shifted to remedies that provide boosts to our health (deal with bad backs, headaches, other symptoms of stress and fatigue) and yet do not tackle the root causes of those issues, namely, being overworked (2015: 40, Thanem and Elraz, 2022). Shepard walks among bodies, that are to be processed and genetically changed to create reapers, fed into a wider war machine, set upon ending civilizations. The bodies have been wasted, just as in *The Last of Us* they lie to decay in ruined office buildings.

The Transhumanist Experience in a Wasted Land

Zombies are a visceral wasted body, but *The Last of Us* takes a different approach to most media of this post-2000s era. For one thing, it is a zombie game where the player must confront the monstrosity externalized. The fight against the wasted zombie, is a fight to resist the zombie path into an all-consuming machine, even though arguably those uninfected are replicating that sole-minded life as humans compete for limited resources. This competition becomes their sole focus for survival. The zombie is a confrontation with the truth, at a time when humans could choose to build up any system, but fall utterly and hopelessly back to the old systems that failed. In these narratives, the occupants cannot envision a life beyond capitalism (Le Guin cited in Hachadourian 2015: n.pag). Even as populations are laid to waste, the survivors (often fighting the wasted bodies) are trying to replicate the old systems that failed. Additionally, the story shows the permanent wasting of bodies, years after the start of the infectious outbreak.

The difference with this and other zombie stories is subtle. There is a sense of rebuilding, as seen by the communities starting to spring up such as Jackson, the one Joel's brother is affiliated with. The realization that under many post-apocalypse scenarios, the only real lasting change is the erasure of technology that supported systems (such as healthcare) and not a fundamental shift in economics, competition for land, space, or food. The climate catastrophe is ultimately a greater tragedy for the knowledge that humanity would replicate the same mistakes, and is already beginning to do so, even with so little left from the previous system. The relationship between places and bodies fluctuates, particularly when migrants seek to leave spaces and face different, evolving, and unstable systems and rights. Throughout the majority of *The Last of Us*, the player is a wanderer, trying to navigate safely from space to space. The player's experience is a migrational one for the most part in *The Last of Us*, travelling from ruined town to town, seeing what has been lost and also witnessing the intense competition for scarce resources in almost every space that Joel and Ellie arrive at.

> When we think about towns and cities and the spaces there in which we live, we bias our thoughts about permanence, the people and things that are always there. We often fail to perceive how these permanent spaces appear to the temporary visitor just passing through, maybe stopping for a time, but ultimately moving on to another destination. We fail to consider fully how a migratory population builds and dwells for a short time within a permanent space. For this group of 'transhumans,' nothing is permanent, and these spaces and locations become commodities to use and leave behind. (Sennett 2018: 104)

Sennett's work on both the concepts of environmental building, and on Heidegger's legacy of the philosophies of building and dwelling, provides revelations about the nature—or even unnaturalness—of the spaces we construct, occupy, and abandon. For much can be revealed in the need, desire, or decision to abandon a place or space. In *The Last of Us* these temporary spaces are akin to tombs, where bodies and people were left except for the startling fact that no tribute could be made to them. No monuments erected, and sometimes the only clue as to who the dead are comes from little notes or scraps of news clippings left around for the player to stumble upon. Despite entire villages, towns and cities being wasted and left derelict as a symbol of the recent bloody history that humanity has experienced, the microhistory has been lost as people have been consumed by the virus, and the infected become workers set upon infecting more people to join the ranks of the clickers.

As Sennett refers to those who travel through places as transhumans, there is, of course, another kind of transhuman—those who push the boundaries of the biologically and technologically possible to become hybrid beings, those which Donna Haraway dubbed "cyborgs" (2006). *Mass Effect* examined a particular type of cyborg: The zombie cyborg. While Haraway presented a utopian ideal of cyborgs, envisioning that they would break down gender binaries, barriers and connect us not just as individuals but as networked communities, the reapers followed that path but used that to ultimately make all upgraded humans (husks) fall under their control, and abolished any concept of autonomy. The focus on the networked body gave more attention to the network, and communications technology than it did the biological form it needed as its basis.

Transhumanist and feminist scholar, Alaimo contested this, claiming that feminist studies had largely ignored the materiality of the cyborg—including its flesh, bones and basic biology in favor of focusing on the technological, claiming "the critical reception of the cyborg as technological but not biological insinuates a transcendent cyberhumanism that shakes off worldly entanglements" (Alaimo 2020: 7). Additionally, Alaimo asserted that the body has been delineated from its surrounding environment, as a focus has shifted toward the fetishism of genes and our bio-structural makeup (2020: 106). Environmental inheritance is another factor of familial structure and legacy (Alaimo 2020). *Mass Effect* thematically would support the Haraway view; the cyborgs become so disinvested and disembodied from their biological forms that they can no longer differentiate between different individuals for every living organic form is a potential new recruit to the reaper work force. However, this is a dystopian view of cyborgs, not as liberated, but rather nothing more than a cog in the reaper war machine, certainly not a tone in-line with Haraway's work. The technology that humans use to enhance life becomes weakness that the reapers target for their own control and to use against humanity and the other organic alien races.

No one character in *The Last of Us* fits the bill of the cyborg; the pandemic happened so swiftly and so severely that there are barely working lights and windmills even in the most secure towns. Communication technology is simply non-existent, which is why Joel and Ellie spend the entire first game trying to locate the Fireflies. Biology and matter cannot be escaped when Ellie had previously been bitten and yet is the only known human not to have "turned." The future of humanity can be mapped through Ellie's resistant body, as the potential for a cure dwells within her. Ellie therefore is biologically significant to all those around her, and yet it is not the thing that defines her, mostly because very few characters know of Ellie's immune status. People work and operate together as comrades, working together against the hordes of infected and clickers. In this stripped-down society, the body is trapped by the worldly entanglements that revolve merely around infected status, a status that determines life expectancy, support, access to communities and access to mutual aid and solidarity, such as sharing space, food, and vital equipment.

The term virus not only refers to threatening life through infection but was long associated with changing or transformation. Coupled with this is the idea of changing communities—those who do not succumb to the disease, inevitably build immunity (Vint 2013: 139). The sense of community however, is not allowed to build in *The Last of Us* as no other known human is immune. Ellie does not become the focal point of a community, but she embodies hope in the material sense as her biology can be used to formulate a cure. Yet, the process is inevitably not quite so utopian. Ellie submits to being experimented upon but does not realize that any cure taken from her body, would result in her death. Ellie's life, and potential and individuality may be wasted for what is seen as the common good by the Fireflies, in a procedure that could have potentially saved humanity from the virus plaguing the world. Joel, after forming a strong connection to Ellie, cannot allow this scenario after the death of his daughter, Sarah, at the start of the pandemic. The player has no choice but to continue on as Joel works his way through the hospital, slaughtering the Fireflies. In the final room, Joel meets an unconscious Ellie and will take her from the Fireflies. Joel shoots the lead surgeon, but the player has the choice whether to also execute the other remaining healthcare workers who are defenseless. The base narrative that zombie stories often descend to—how far would you go to protect your loved ones in a lawless society—is set against the backdrop of wasting.

The reclaimed Firefly hospital becomes littered with wasted bodies, not infected, but ultimately left for dead due to the pandemic and the tremendous sociological and ideological stakes it has created. "Body" scholars' have long identified the competition within neoliberal capital as a cause of wasting bodies. Klinenberg investigated the "political sociology of bodies" (2002: 123) in relation to excess deaths during the Chicago heatwave of 1995. The surge in unexpected deaths was exacerbated by the entrenched inequality of Chicago, afflicting the poor, disabled and Black communities—such as the south and south-west sides of the city—made up of African Americans (2002: 125). Klinenberg goes so far to argue that bodies were so visible with their wasting as morgues could not accept entry that the public did not know how to comprehend this mass death. As the rich fled the city to their beach homes, those without privilege were trapped within the neoliberal system that only served to put the local population in danger.

In The Last of Us, people still compete for safety, land and opportunity just as they do under a sustained technological capitalism, but the bodies are left wasted over a struggle over biological autonomy—who has greater rights, Ellie or the Firefly group? Ultimately, the players are left to dissect the horror of what they have encountered, even if they ideologically align with Joel, as the wasting of bodies continues in a pandemic that has seen the destruction of all infrastructure, and now the last remaining healthcare facility. Ellie is saved, once more from the fatality of the pandemic, the one true posthuman who has ascended the biological limits infecting populations around her. Christie and Lauro (2011) argue that what creates the discomfort in posthuman horror, is the realization that humanity is out of place. Joel, Tess, Sam, Henry, and the myriad of other characters that the player meets, are all one step behind Ellie who has already been guaranteed a future and safety away from a virus that the rest must all grapple with, and that creates the desperate confrontations over scarce resources (resources still in demand, but now with no infrastructure to create them).

This is an overt fear that Shepard is informed of repeatedly by the reapers, as the Commander is told by Harbinger, the main reaper protagonist, and his fleet that it is in humanity's interests for the reapers to harvest the genetic material of trillions of lives so that they may become preserved in perfected reapers, and the chaos (of war) will end. This links with Wasson's medial humanities research into the organ (legal and illegal) trades, after focusing upon the motif of clones in science fiction, using Ishiguro's *Never Let Me Go* as inspiration (2011: 75). Wasson compared clones to spare parts, and found that under constructs of wealth and power, organs pass "from marginalised to powerful," from poor countries to the West (2011: 80) exhibiting the pressures for the poor to waste their own bodies (sometimes without informed consent) in the hopes of a better life (2011: 79). In *Mass Effect* the player can indeed allow the harvesting to continue and see organic bodies as spare parts for the reapers, but the game ends with nothing left of the Milky Way universe, and with the death of communities the player had come to build. Ultimately, the attention of both games shifts to the bodily rights of the people around us, whether systems matter beyond individuals, and whether bodily organs are controlled by the state/acting-governments or by the individuals who carry them.

Conclusion

In a climate catastrophe, it is often the world around us that become the focus of multimedia (and phenomenological) experiences, but a climate disaster (or ruin) is inextricably linked with the wasting of bodies. *The Last of Us* and *Mass Effect* show the two specific ways in which bodies are wasted: Premature death and loss of hope from the loss of structures that populations become accustomed to and dependent on, and the desire for the wasting of bodies through theft of genetic material (the Fireflies and the reapers).

Materially, the neoliberal structure of extracting the most resources from the workforce and certain demographics becomes futile, as the systems crumble under an apocalypse. Those forces governing society no longer remain, however, it is the shadow lingering over the rebuild of any community. It is what was lost, and what people strive to achieve again through bioethical memories of the system they had before it all went wrong. *The Last of Us* demonstrates this as communities fight for limited resources, and strangers are more likely to be competitors and foes rather than friends in the struggle to survive with the same dependencies on processed foods where scavenging becomes essential but there is no way to maintain that infrastructure.

Yet, *Mass Effect* portrays an enemy bringing impending accelerated climate disaster that is a stronger force of neoliberal bio-extraction. The reapers have found a more effective way of securing labor and biomaterials. While the current structure crumbles around humanity, in its ruins isn't waste, but instead a more relentless pursuit of ensuring that a labor force is guaranteed. The systems in place become redundant, only because humanity transforms into reaper agents once their genetic material is taken. The move away from neoliberal structures such as the council, only comes as populations are consumed by a greater power.

The body is wasted, and the husk is discarded under any climate catastrophe, which is inevitably created and caused by resource driven consumerist capitalism. Just as plastics are wasted, used, and clutter the landscape, so too will bodies when the world can no longer sustain intense populations concentrated particularly in certain areas, or those populations that are unfortunate enough—due to circumstance of birth into poverty—to be at the pressure points of changing landscapes, with few healthcare systems or social structures to protect from the worst effects of a changing world. Just as it was in *Mass Effect* where humanity was left to suffer the invasion of the reapers first (with the batarians) simply because wealthier races—the turians, salarians, and asari—wanted the time to protect their own wealth and resources, so those with less power and capital were sacrificed in order for the most powerful to protect their assets.

Bio-capital becomes a commodity to be fought over. Immunity, resilience, biological advantages—our genetics, become a battleground under biopunk stories. The wasting of bodies is a last Foucauldian vision, when structures come under strain and there is competition to become the most perfected neoliberal state, accelerated by the tension and anxiety of a landscape and climate crumbling around us.

Heidegger noted that it was difficult to dwell and this creates an ontological insecurity (Sennett 2018: 124) as people move from place to place, struggling to form attachments as society progresses and places fundamentally change. There is a huge contrast between the two science fiction stories, as *Mass Effect* shows the anxiety through homes being ruined by the reapers in the most overt and destructive ways while *The Last of Us* depicts the longing for security in a stable place, that may be possible but is something Ellie shuns in order not to feel wasted, so that she may contribute to the cure for humanity. The place then, while inextricably linked to mental health, is also tied to the physical body. As reapers lay waste to homes to easily extract human genetic material, and the people of the southern states are vulnerable without four high walls around to protect them from both clickers and human predators. The body border does not end at biology but begins with the physical world around us:

The wasting of bodies can only happen if security and structures are left to dilapidate under late-capitalism.

References

Alaimo, Stacey. (2020), *Bodily Natures: Science, Environment, and The Material Self*, Indianapolis: Indiana University Press.

Bocock. Robert. (1993), *Key Ideas: Consumption*, London and New York: Routledge.

Christie, Deborah, & Lauro, Sarah Juliet (2011), *Better Off Dead : The Evolution of the Zombie as Post-Human*, United States of America: Fordham University Press, Ebook.

Cole, Tom, and Marco Gillies (2019), 'Thinking and Doing: Challenge, Agency, and the Eudaimonic Experience in Video Games', *Games and Culture*, 16:2, pp. 187–207, https://doi.org/10.1177/1555412019881536. Accessed 20th November 2022.

Comaroff, Jean. and Comaroff John (2013), 'Alien-Nation: Zombies, Immigrants, and Millennial Capitalism' in S. J. Lauro (eds), *Zombie Theory: A Reader*, Minneapolis: University of Minnesota Press, pp.137-156.

Cooper, Melinda. (2008) *Life as Surplus: Biotechnology and Capitalism in the Neoliberal Era*, Seattle and London: University of Washington Press.

Foucault, Michel. (1976), *L'Histoire de la sexualité* (*The History of Sexuality*), New York: Random House.

Foucault, Michel. (1995), *Discipline and Punish: The Birth of the Prison*, (trans A. Sheridan) New York: Vintage Books.

Hachadourian, Araz. (2015), 'Ursula K Le Guin calls on fantasy and sci fi writers to envision alternatives to capitalism', *Open Democracy*, 12 June 2015, https://www.opendemocracy.net/en/transformation/ursula-k-leguin-calls-on-fantasy-and-sci-fi-writers-to-envision-alt/ Accessed 14 March 2022

Haraway, Donna. (2006), *A Cyborg Manifesto: Science, Technology, and Socialist-Feminism in the Late 20th Century*, Weiss, J., Nolan, J., Hunsinger, J., Trifonas, P. eds. Dordrecht, Springer Netherlands.

Klinenberg, Eric (2002) 'Bodies That Don't Matter: Death and Dereliction in Chicago', in N. Scheper-Hughes and L. Wacquant (eds) *Commodifying Bodies*, London: SAGE Publications, pp. 121–135

Kristensen, Kasper Simo. (2013), 'Michel Foucault on Bio-Power and Biopolitics.' University of Helsinki.

Kristeva, Julia. (1982), *Powers of Horror: An Essay on Abjection*, (trans. L. S. Roudiez), New York: Columbia University Press.

Levina, Marina. (2009), 'The Regulation and Discipline in the Genomic Age: A Consideration of Differences between Genetics and Eugenics.' In S. Binkley and J. Capetillo (eds) *A Foucault for the 21st Century" Governmentality,*

Biopolitics and Discipline in the New Millenium, Cambridge: Cambridge Scholars Publishing, Web.

Mateas, Michael. (2001), 'A Preliminary Poetics for Interactive Drama and Games', *Digital Creativity*, 12:3, pp. 140–152., https://doi.org/10.1076/digc.12.3.140.3224. Accessed 20th November 2022.

Meek, Lisa. (2022), 'Dwelling in Animal Crossing', *Games, Culture and Identity: A MultiPlay Conference*, University of Sunderland, January 12 https://www.youtube.com/watch?v=KsJue7JJzyg&list=PLdIhov-l8FHVf0GQl-EWEyADvkfwvUEUa&index=4

Mitchell, David. T., Sharon L. Snyder (2015), *The Biopolitics of Disability: Neoliberalism, Ablenationalism, and Peripheral Embodiment*, United States of America: University of Michigan.

Monbiot, George. (2016), 'Neoliberalism – the ideology at the root of all our problems', *The Guardian*, 15 April 2016, https://www.theguardian.com/books/2016/apr/15/neoliberalism-ideology-problem-george-monbiot Accessed [4] April 2022

Muriel Daniel., Crawford, Garry. (2018), *Video Games As Culture: Considering the Role and Importance of Video Games in Contemporary Society*, Oxon: Routledge.

Poole, Steven (20177), 'Why Video Games Are Obsessed with the Apocalypse', *BBC*, 15 August 2017, https://www.bbc.com/future/article/20170815-why-video-games-are-obsessed-with-the-apocalypse?isappinstalled=0. Accessed 16 ovember 2022.

Sennett, Richard. (2018), *Building and Dwelling*, London: Penguin Books.

Schmeink, 2016 Schmeink, Lars. (2016), *Biopunk Dystopias Genetic Engineering, Society and Science Fiction*, Liverpool: Liverpool University Press.

Tanenbaum, Karen, and Theresa J Tanenbaum. (2010), 'Commitment to Meaning: A Reframing of Agency in Games', *EScholarship*, 25 January, University of California, https://escholarship.org/uc/item/6f49r74n Accessed 10 November 2022.

Thanem, Torkild, and Hadar Elraz. (2022) 'From Stress to Resistance: Challenging the Capitalist Underpinnings of Mental Unhealth in Work and Organizations', *International Journal of Management Reviews*, 24: 4, pp. 577–598., https://doi.org/10.1111/ijmr.12293. Accessed 27 November 2022.

Vint, Sherryl. (2013), 'Abject Posthumanism: Neoliberalism, biopolitics, and zombies', in M. Levina (eds) *Monster culture in the 21st century: a reader*, London: Bloomsbury, pp.132–146

Wasson, Sara. (2011) 'A Butcher's Shop where the Meat Still Moved': Gothic Doubles, Organ Harvesting and Human Cloning', in S. Wasson and E. Alder (eds) *Gothic Science Fiction 1980–2010*, Liverpool: Liverpool University Press, pp. 73–87

Wellman, Barry, et al. (2006), 'The Social Affordances of the Internet for Networked Individualism', *Journal of Computer-Mediated Communication*, 8:3, https://doi.org/10.1111/j.1083-6101.2003.tb00216.x. Accessed 20 November 2022.

PART IV

New Theoretical Horizons

CHAPTER 14

Three Digressions (on the Way to Eco-Aesthetic Politics)

Seán Cubitt

A Digression on Mermaids

One day in 1842, Moses Kimball, proprietor of the Boston Museum, purchased a mermaid. He bought the mummified creature, about two or three feet long, from the son of a sea captain named Eades who had bought it in the back streets of Batavia after selling his share in a trading ship to raise the funds. It seems the mermaid was a masterpiece of creative taxidermy, including parts of shark, orangutan, monkeys, dogs, and fish skin, and that there was quite a trade in 'feejee mermaids' in the Dutch East Indies at the time. Eades showed it round London to moderate success, and his son had a go at showing it at his lodgings but lost interest and offered it to Kimball, who enlisted Phineas Taylor Barnum, already known as New York's go-to publicist. Barnum set up his lawyer, Levi Lyman, to play the part of an apparently reluctant English biologist who

S. Cubitt (✉)
Screen and Cultural Studies, School of Culture and Communication, University of Melbourne, Parkville, VIC, Australia
e-mail: seancubitt2017@gmail.com; scubitt@unimelb.edu.au

D. Binns and R. Najdowski (eds.), *Confronting the Climate Crisis*, Palgrave Studies in Media and Environmental Communication,
https://doi.org/10.1007/978-3-031-89606-4_14

would deliver lectures on it and on his own adventurous but wholly fictitious life. After a year, Barnum sent the creature on a tour of the American South in the hands of his uncle who, however, lacked Lyman's gift for ballyhoo. Undeterred, Barnum took the mermaid to London with some success, before installing it at his American Museum where it was most probably lost in a fire (I owe these details to Bondeson 1999).

Spectacular quasi-scientific entertainments were important venues for popular imaginings of the Orient (Japanese mermaid stories go back at least to the tenth century, with possible sources in China of the Warring States period: Chaiklin [2010]). There was a kind of contractual nod-and-wink linking pseudo-intellectual showmanship with a newly urbane and knowing crowd who enjoyed the skill of the illusion, as a type of confidence trick that cleverly jumbled class barriers to mix quasi-scientific, quasi-religious, quasi-philosophical interests, long features of carnival, into the history of popular amusements (Alderson 1992; Rittvo 1997; Cook 2001).

If the mermaid of legend was the favoured fantasy of water-weary seamen sick of rum, sodomy, and the lash, then the con-trick monkey-fish is the truth of that hallucinatory wish-fulfilment. Its first Western owner, a Captain Eades was perhaps the last to see beyond the nasty relic to the mythic creature that it purported once to have been. The fascination of the relic lies at this hinge between the dream of beauty and the actuality of ownership and sale.

Advertised as the 'Feejee Mermaid', Barnum's orientalist fantasia ballyhooed the glamour of the Spice Islands but in a perverse doubling of hallucinated loveliness with no less hallucinatory disgust at the mire of sex and putrefaction. This precisely fantastic doubling of desire and death is the last illusion, one that pretends 'finally' to have transcended sexuality and sensuality along with dream. The obverse of mythic mermaids, the Feejee Mermaid made explicit a puritan disgust at heterosexual congress through the conjunction of the lecherous monkey and seafood. The mummification is, at first glance, the anomaly: the caesura of preservation between the rhymes of sex with putrefaction.

If we could possess the object of our desire, we would cease desiring, and that never happens. Barnum's Feejee mermaid is of course the bestial imago of desire in a decaying mirror. But mummification turns its parody of love into fantastic fiction when it pretends to show what cannot occur: the end of desire. The saddest thing is that in some mediaeval carnival, it might have lived up to its promise, showing that sexual desire transcends even death. But in Barnum's carny, the dime-show freak is no longer capable of the mythic and fantastic reversals of carnival. It is instead assimilated into commodity exchange, a service provided.

A mermaid swims between two worlds: the business-like shipboard world of sailors, the vast majority of whom, like Captain Eades, couldn't swim, and the sea-caves of the deep where treasure lay piled and fantasies of sexual play contradicted the otherwise fatal nature of water. In the imaginary ocean, desire and death combine. There was, especially for London audiences, another female creature perched on a rock on the shore, singing songs that lured young men to their deaths. Staunch with helmet, buckler, trident, and shield, regal Britannia, allegorical personification of Empire, comely negation of the sensuous, was the imperial antithesis of mermaids, overwhelming desire with an equally unending duty.

Compare Gertie MacDowell, perched on a rock on the beach at Sandycove in *Ulysses*, object of Bloom's prurient desire until at last she reveals her limp, her imperfection, her unworthiness. The terrible bathos of Joyce's epic, the reduction to banality, makes of Gertie a Feejee siren; but it also celebrates the everyday as the height of epic truth. Even under Bloom's demeaning gaze, Gertie does transfix and is transfigured. Barnum's sideshow cannot pull off that second trick. He lacked Joyce's ability to see the radiant legend beneath the naked skull.

Instead, Barnum's Feejee mermaid secularises three sacred systems. One, the ancient belief in water spirits; two, the millennia-old fantasies of sailors; and three, the seizure of both in Britannia as symbol of imperial power. Animist belief in local spirits recognised in desire what we would now call 'affect' where rivers and seas longed for the bodies of sailors as much as sailors yearned for fantastic carnal bliss. Sailors' mermaids emerged from the repressive discipline of ships, a desire perverted towards possession, in a fantasy that also dreamed of being possessed. This second desire was exclusively human. Britannia, the third, was always a sublimation of desire, power as austerity, alone on an anonymous and depopulated shore. But even this sublimated mermaid hauled itself out of the economic calculus that funded it, a desire converted into bond among men socialised by the secret pledge of duty that they kept private in each doughty heart.

That imperial honour that Kipling's (1899) Victorian troopers keep, a flag folded and hidden, reassembled the ecological flow of desire through humans, nature and technologies like ships, the doubled fantasy of love and death, and a commons bound by shared desire, albeit in service to a colonial fantasy of omnipotence. Fantasy mermaids mediated the incompossible worlds of sea and ship; Feejee mermaids mediated the East to Westerners, stemming at origin from a founding alienation experienced as the oscillation between desire and disgust. The equally profound alienations of the commodity form, in which navigation, Orient and desire met their cynical equation, derives from its failure to transcend the condition it articulates as sales pitch.

Stripped down to a cash transaction, the sexual act is no more than a pretence that purchase can bring desire to an end. But the only end of desire, which is by definition perpetual, is death: and that is what the Feejee mermaid pictured. As carnivalesque parody of Britannia and squalid reveal of prostitution, Barnum's mermaid was an act of resistance. But its cynical conjuncture of con-trick with the knowing spectatorship of knowing dupes simultaneously denied the utopian possibilities of desire in a new economy where common knowledge could become private property. It should have stopped once it had denounced Romantic individualism and imperial honour, the perverse offspring of repression and sublimation. Instead, Barnum's commodification of desire annihilates one by one desire's transcendence of repression, its capacity to form a commons, its ability to transcend death, and the play of desire through bodies that are always wholly immersed in planetary ecologies. The greatest sacrilege of the Feejee mermaid is that it reduces affect, the world's desire, to a grotesque mummy that ultimately only expresses a regime based on individual psychological and legal possession.

A Digression on Angels

Angels appear; that is their duty and their glory. Their appearance is *here*: in the human world. But their *being*, as creatures of other worlds, is not exclusively in the same place as their appearing. The angel is a creature of the threshold. There have always been gods of doorways and gateways: angels of the door, for whom every door is potentially the corridor between orders of being. For St Thomas Aquinas, 'the angelic doctor' of the Thirteenth Century,

> the perfection of the universe requires that there should be intellectual creatures. Now to understand cannot be the action of a body, nor of any corporeal power.... Hence the perfection of the universe requires the existence of an incorporeal creature. (1920: *Summa Theologica* 1a 50, 1)

The obvious secular response is that the incommensurable worlds of goods, services, labour and consumption *appear* without the action of 'any corporeal power', nominally through the medium of money, the universal commodity, but effectively as information, now that notes and coins are residual physical markers of actual electronic wealth, which itself is not the zeroes and ones that compose it or the quantum events of the electronic infrastructure but a purely intellectual agency without human, technological or ecological ground. In a Talmudic variant, according to Benjamin (1999: 457), there are angels who 'are at each moment created anew in countless throngs, and who, once they have raised their voices before God, cease and pass into nothingness'. Like money, these angels have neither past not future, and like money and quantum particles, the ground of reality evaporates as anything other than witness to their momentary appearance.

Aggelos, angels, are messengers. Transcribing Aquinas into the mathematical theory of communication, angels are signal, intellectual and incorporeal, and as many of them can dance on a silicon chip as God wants. Whatever hinders signal is noise; which is communication theory's name for matter—the matter of the external world or of the communication medium itself interfering with the pure abstraction of signal. If angels are wholly spiritual they are wholly signal. But if angels are media, their existence is material, which implies that they are noise intersecting our instruments of communication.

Angels announce, and when and what they announce matters, in all senses of the word. In his theses 'On the Concept of History' Walter Benjamin (2003) describes the *Angelus Novus* being blown backwards out of Paradise by the storms of progress. We might say it caught in its wings the blast of the atom bomb. Indeed, the Bomb itself is an angel, a messenger between worlds, between orders of being. Arendt describes humanity's entry into history through the linear stories of heroic deeds, each of which is the irruption of the extraordinary event into the natural, cyclical order. The bomb interrupts this linear structure of history with the force of a third order of existence, neither natural nor human.

An atomic blast is the energy released when the Absolute enters time to make itself manifest. Fatally beyond comprehension, the bomb is an angel that exceeds the capabilities of those that it appears to: the angel Azrael that takes souls out of this world, the angel of death. What we call history is the result of this fatal arrival. The angel of destruction, this interruption that ends history and creates whatever it is that we now inhabit, appears as a new truth. It enforces the non-perceptual actuality of physics on a population who had thought themselves formally and finally divorced from the natural world. The revenge of Being on appearance. Like every angel, the bomb arrived as a stranger—and continues to arrive, even in a time when we have forgotten that it dwells among us, the invisible trigger that turns alienation into annihilation.

The world is for us no longer historical or natural. The global system that has bootstrapped itself out of the swamp to become pure appearance, cash divorced from assets, consumption divorced from ecology. The angelic bomb and the atomic sublime that attends it as a new and terrifyingly nihilistic aesthetic inhabit the interstices of the simulation enacted everyday by finance capital, an economy of increasingly remote and arbitrary derivatives that constantly reformulate the grounds of their own existence, the market. As interruption, the angelic event, that has not ceased to undercut the everyday with its permanent alternative of annihilation, proves that this simulation, which must aspire to totality, can never be total because it is always dependent on the world it boasted it had bootstrapped itself out of.

In the bomb's dark interchange of projection and introjection, the world's and our desire to be annihilated articulates with the radical incompletion and totality of the sexuality of angels. In the Byzantine Church, since Christ had referred to the angels in heaven 'neither marrying nor being given in marriage', it became customary to regard eunuchs as angelic (McGuckin 2017: 850). In related traditions, hermaphrodites were angelic, having the parts of both sexes and therefore being whole, whereas the sexed population are not (Origen 1885: Book 1, part 5.1). The inference of both traditions is that angels are at once of both genders and neither, lacking and complete. The angel of the threshold faces both towards and away from sexuality, popping in and out of identity as easily as, in the Talmud, they pop in and out of existence (Agamben 2011). Not gendered and, in many traditions, not even asexual, angels persist *between* worlds and orders of being, between reality and symbol. They are neither pole of these binaries, but messengers, relations, mediations between them.

As message, as threshold, as event, the angel is wholly relational. It has no Being of its own. It *is*, to the extent that it 'is' at all, a flowering of unstable and failing presence, perpetually coming into and fading out of being in the gaps between manifestation and witnessing, completing and rejecting the terms of desire, an event that can never wholly appear to a witness who is never entirely reliable. The Angel is not a person but an encounter. The encounter with an angel is the strange moment when we realise that 'I'—as witness—am not identical to myself. The angelic encounter forces a new relation with the world but in doing so blasts open the prison-house of identity, engineering a second encounter of the witness with what previously appeared to be a self. Though it must begin as trauma, the angelic encounter is a departure. The bomb marks a departure. In Bernini's Ecstasy, experience of the sublime is an exit from the world. At Bikini Atoll, like the angel that drove Eve and Adam from the Garden, the angel of the fiery sword drove us out of history into an endless, systemic present. This present—the present of the market most of all—is inhuman because it denies what the departure from eternity into history made apparent: that human time is marked by loss. Loss of ecological, maternal and communal union, felt as profound lack, binds us to a fading past and propels us into the future. The paradox of the angelic encounter's queering of desire in the face of the nuclear sublime—and the equally aesthetic terrors of climate change—is that each of us experiences it alone, yet all of us experience it in common. The overwhelming realisation of our alienation from the world and one another propels us towards one another and a world that has just proven that it is unstable, and therefore capable of change.

Powered by the medium of the word (John's gospel calls it *logos*; the Quran opens when the angel Jibril commands the Prophet to write), the highest archangels and most modest guardian angels know us. Angels assure us that we are known by something that cares.

A Digression on Cats

Like the market, the pandemic knows us but doesn't care. Ana Lily Amirpour's 2015 film *A Girl Walks Home Alone at Night* predates the pandemic by five years and shows us how the *world* might care about humans. Like all vampire movies it evokes blood as medium, unsettles the connotations of menstruation and thus of femininity as lunar and reproductive, thus proximal to nature. The gender-switch—Amirpour's vampire is female—already breaks the code: the Girl (unnamed throughout the film) is un-natural. Beyond the obvious allegories—perfect in their ambiguity—there lies the challenge of the film's ending and of the role of one of its principle characters. The plot is a love story and ends as the young male character Arash and the Girl drive away from Bad City, the petrol town they inhabit and that inhabits them, shot in chiaroscuro black and white in another oil-economy town in Southern California, a diasporan doppelganger that displaces the Farsi dialogue against evocations of American youth culture (US rock music, James Dean car, clothes and poses, US stars on poster and, in a marvellous sequence, the Girl skateboarding in a chador). After driving for a while, they park up. Arash gets out. He seems to be making some kind of decision. He gets back into the car and drives. The final shot shows taillights receding. But which way are they going—escaping the city or returning to it?

The love story involves a boy and the Girl, but elsewhere the film plays with all the roles. With flamboyant make-up, Arash dressed as Dracula dances with the Girl at a costume party; earlier she has been treated to a would-be erotic dance by a pimp who she bites and kills. In another unexplained insert sequence, a trans character we have been introduced to briefly as a streetwalker controlled by the now-dead pimp dances alone on a rooftop with a helium balloon for partner. This mirror of the pimp's attempted, assertive, almost bullying seduction routine is more joyous, celebratory, and asexual, as the core love story also appears to be. The play of roles, genders, sexes and asexualities elevates the movie from realist narrative towards some kind of symbolism. But of what? Perhaps the question is better asked, of the rooftop trans dance, and of another sequence of the Girl dancing alone: who is watching? In many films the answer would be 'the cinema', what Gaudreault (2009: 72–80) calls 'le grand monstrateur', the One that shows us the events on screen. But *A Girl Walks Home* presents an alternative to this self-reflexive analogy of the market's self-sufficient Presence.

In the opening scene, Arash collects a feral cat from the outskirts of town. The cat accompanies him through the story, often caught in close up, watching events unfold. At moments it enters into the plot, first being sexed by the pimp (who calls it Mr Cat), later by Arash's junkie father who accuses it of 'having your mother's eyes'—another gender uncertainty.

The cat is a cat, as real as any of the actors. In sequences in the car, where it perches between Arash and the Girl, it seems to follow the alternating lines of dialogue, turning its head towards each speaker, perhaps a trained animal, an actor, but no less a cat than Arash is less human for being a performer. At the same time, obscurely, and in ways that may never yield up a satisfying resolution, it seems to have the role of watching, witnessing, but perhaps least and most specifically of seeing. It is a principle of Merleau-Ponty's (1968) late phenomenology that in order to see, we make some kind of compact with the world that we must also be seen, inscribed in the field of the visible. The cat then operates in the film as the world's view of the action, an uncanny gaze beyond the suturing of male desire into the order of continuity editing (Mulvey 1975), even as the turns of its head suggest a classical shot-reverse-shot structure that, however, replaces the implicit masculinity of the *grand monstrateur* with a feline look, dispassionate, almost objective except that it too is an object of the camera's gaze and therefore not free to be an impartial frame or framer of actions and objects.

The grounding ambiguity, the non-identity of the cat (simultaneously cat, symbol, observer, actor, function of the film, emblem of the world that sees rooftop dances and automobile romances, impartial, beyond the film but at risk at least twice in the narrative ...) undermines any ordinary reading of *A Girl Walks Home Alone*. Rippling between the anthropomorphisms (Mr Cat, mother) and animal embodiment, between feral and domesticated, between subject and object of the film's partition of looking, the cat is an unsettled and unsettling presence of un-nature, an absence of either nature (dispelled from Bad City) or reason (or why would it matter if a cat sees anything?). The cat that persists throughout the movie is, rather precisely, nothing. A mote in the eye, a patch of invisibility in the omnivoyant world of cinema.

The film was shot—and most frequently shown, in the age of DCP (digital content package) projection, optical discs and streaming media—digitally. A line in Godard's film *Le Petit soldat* (1963) echoes through the history of analog cinema: 'Photography is truth. Film is truth twenty-four times a second'. Almost certainly he wanted to say something about the presence of the image, either as witness to what it observes, or in the sense of a truth-statement, a form of language capable of declaring that some state of affairs is or was indeed the case. But there is another way to read this: twenty four times a second the shutter comes down, in the camera and the projector, to hide the film transport mechanism.

Nothing appears on screen. And what is nothing? For Frege (1953), as for Aristotle, everything that exists is identical to itself: this cup is this cup, not that cup or that tree. What then is zero, which by definition does not exist? It is the name of the non-identical. The darkness we plunge into, and during which we are, again in the full technical senses of the word, unconscious, is this zero: the integral non-identity of successive frames that their presence as images emerges from. The multiple non-identities of the cat may as well serve as on-screen personification (animation) of this precisely invisible generation.

Except that this is not film in the sense Godard spoke of in 1963, long before he moved to video.

In video projection, there is no shutter. Instead, each frame is scanned, from top left to bottom right, the screen illuminating and fading just slowly enough to maintain the optical illusion of a whole image, but swiftly enough to clear the way for the next image to emerge. In video, unlike film, after the first exposure, there is never a whole image. But equally there is never a complete darkness, only this pulse of emerging and fading. This rhythmic infrastructure, which only operates on the condition of being invisible to human eyes (but which cats can see, hence their general lack of interest in screens) tells us something odd about the digital image (equally true of stills, which are also scanned repeatedly): in digital media there is neither presence nor absence, no zero and no one. We tell ourselves that the truth of video is not pictures but code, recusing it from being anything more than a statement about statements, withdrawing from the world into aesthetic autonomy. Yet we know this is impossible: we know it depends on code administered by global standards for image transport, running on devices fabricated in some plant, made of metals and plastics transported, refined and managed, derived from some mineral resource. And we recognise the activities involved: actors, locations, lights, microphones, production crew, machines—and cats. We often note the ontological uncertainty of digital video, based on how malleable it is. I like to think of this as a state between being and representing, the subjunctive mood of 'could' or 'should' or 'might'. Video's uncertain status echoes both the unstable status of the cat (animal-symbol, observer-observed etcetera) *and* the quantum uncertainty of electrical charge, which can never achieve completion, and never decline to zero. The language of zeroes and ones, presence and absence, is a fiction. The cat in *A Girl Walks Home* looks less like a Persian cat and more like Schrödinger's (Trimmer 1980).

Or like the Cheshire Cat in Lewis Carroll's *Alice in Wonderland* that fades away leaving only its grin, and even more like Kipling's cat who walked by itself, the scale of difference between one image and the next, between one form of unconsciousness and the next, is never settled. For the viewer, watching *Casablanca* on film—when we are unconscious for around half of the movie's running time—or viewing *A Girl Walks Home* on video—when we are semi-conscious all of the time—hardly feels different. But we are no longer unconscious in the same way: not plunged into non-identity, but swept into the rhythm of an alien pulse. Losing analog cinema's absences, its non-identity, we enter electronic space where image and darkness are no longer fundamentally separate states. Appearing and disappearing are our ontology. We are no longer simply inscribed into the visible: we are constantly being marked as visible and equally constantly being erased. Domestic and wild, loved and despised, the witch's familiar and the ecologist's enemy: the privileged figure of this familiar stranger, this strange familiar, is surely the cat.

On the Way to Eco-Aesthetic Politics

Any sufficiently advanced technology is indistinguishable from magic, as Arthur C Clarke observed. To understand the power of the sublime—orientalist, spiritual, technological, animal—we need to understand the way it overcomes not only the senses but the mind. Sublimity generates ignorance, which implies its obverse: where knowledge ends, wonder begins. Any system—including the global market economy, the science we evoke when we talk about 'trusting the science' and many variants of environmental thinking and planning—has to act as if it were complete and coherent. To achieve this completion, a system must actively exclude what economists call 'externalities' and sociologists call 'environments'. Whatever the system embraces it treats as signal; whatever it excludes is only noise. Pushed out beyond the limit of knowledge, wonder is our reaction to those interruptions of the system when psychic, spiritual or worldly ecologies, excluded from the system, force their explosive way back in and demonstrate the fallibility, incompletion and incoherence of the system itself.

The perverse language of information capital reduces knowledge to private property, to property as privation, even as it shrivels the human component of intellect by declaring any corporation a legally constituted person for the purposes of ownership in ideas. The privation of property regimes excludes reciprocity with the world, instituting a mode of knowledge ripe for destruction when the world bursts back in. The perverse language of 'data' and 'givens' that masks the colonial exploitation of reality by the information regime also perverts desire, first by abstracting it from the common flow, and consequently discovering, in the perverse desire to own, a death drive compelled to seek out the vile as if it were the truth of every excluded ecology. For information capital, magic and wonder are illusions created by the absence of knowledge, now defined as intellectual property. But humans are not as divorced from the natural world as capital pretends. Still thoroughly ecological beings, we feel oppression as repression, and experience our alienation worst when the repressed returns as worms and putrefaction. Only a photographer could truly complete the sacrilege against reciprocity, offering us, today, the photographic data of a historic hoax. The world offers to share, but we *take*—a photograph in this case. We call it data, and we 'capture' it without reciprocity. That mad oxymoron 'intellectual property' is all we need to recognise information capital's imperial-colonial epistemology: the world as raw material for extraction, processing, and property. Against this privative mode of knowing we need the ambivalence of a cat's gaze, at once knowing—canny—and uncanny, un-knowing.

Angels keep returning because, despite its triumph, the market, and every information system, is incomplete. Because the human ecology was already ruptured at origin by its entry into history, angels have always been able to pass between the disparate dimensions of the human and its others. These messengers are always strangers, and the obverse is also the case: the stranger is always potentially an angel. This is why neo-nationalist ideologues are so scared of migrants. They are figures of hope not only because they journey in search of better lives but because they renew whoever they meet, cracking open the homogenous present of debt and databases.

What is truly given—not harvested by data-gatherers—such as the arrival of angels and migrants, will not fit into capitalist rationality which demands absolute indifference so that things can be exchanged through the indifferent medium of money. The only differences it accommodates are the zeros and ones of a digital pulse, the difference between the price of copper yesterday and today, or the differences of an ordered hierarchy, such as the one ordering those who pay taxes below those superior beings who do not. These are differences of number, not of kind, and they are incapable of accommodating radical heterogeneity. Capital finds it legitimate to rank billionaires in their orbital heavens but not to greet an angel on Earth. The human becomes commons, as Quijano (2007: 177) has it, by agreeing to the heterogeneity of all reality. Kipling's cat that walked by itself, both domestic and feral, welcomed in and chased away, dependent and independent, is such a creature, that knowingly gives itself to be captured but still evades possession.

These digressions through disgust, terror, and the uncanny began for me with a technical definition. I wanted to say that glitches were evidence of media being true to themselves, revealing their functioning. Glitches are not noise: they are symptoms of noise, where noise is whatever a system elects to exclude. In this sense, a glitch is not even a symptom but the evidence of a system assimilating a symptom into its functioning, which is why we can still hear a melody even on a damaged file. Glitches are like the patina on old paintings, evidence of authenticity: in the end, they only add value to the systems that I once hoped they might disrupt. I've learned, working on this paper, that the more radical the interruption, as in the cases of the Feejee mermaid, the atomic bomb, and the uncanny cat, the less any system—psychological, domestic, economic, technical, scientific, or ecological—can assimilate it unchanged. Any sufficiently repressed reality returns as disgust and terror or magic and wonder, or all four. In the same way that migration challenges the order of the nation-state, in the same way that climate politics challenges the ordered assimilation of states to capital, these encounters are not just disturbances: they demand an entire overhaul of our aesthetics and our concept of politics. Aesthetics can no longer be the administration of taste, any more than politics can be reduced to the administration of public affairs. What breaks both open is *the encounter*, in itself traumatic, which not only destroys an order or determines a new one, but *makes possible*. To make possible is the work of imagination, which is the work of aesthetic politics, ecocritical aesthetic politics: to imagine, beyond the terrible arrival, a world filled with, shaped by, capable of wonder.

References

Agamben, Giorgio (2011). Angels. Translated by Lorenzo Chiesa. *Angelaki: Journal of the Theoretical Humanities* 16(3). 117–123.

Alderson, William T (ed) (1992). *Mermaids, Mummies, and Mastodons*. Washington DC: Association of American Museums.

Aquinas, Saint Thomas (1920). *The Summa Theologiæ of St. Thomas Aquinas*. Second and Revised Edition, Literally translated by Fathers of the English Dominican Province. Imprimatur. Edus. Canonicus Surmont, Vicarius Generalis. Westmonasterii. https://www.newadvent.org/summa/

Benjamin, Walter (1999). Karl Krauss. Translated by Edmund Jephcott. *Selected Writings, Vol. 2, part 2, 1931–1934*. Edited by Michael W Jennings, Howard Eiland and Gary Smith. Bellknap Press / Harvard University Press, Cambridge MA. 433–458.

Benjamin, Walter (2003), 'On the Concept of History' in *Selected Writings, Vol. 4, 1938–1940*. Ed Howard Eiland and Michael W Jennings, Cambridge MA: Belknap Press/Harvard University Press. 389–400.

Bondeson, Jan (1999). *The Feejee Mermaid and Other Essays in Natural and Unnatural History*. Ithaca: Cornell University Press.

Chaiklin, Martha (2010). Simian Amphibians: The Mermaid Trade in Early Modern Japan. *Large and Broad: The Dutch Impact in Early Modern Asia: Essays in Honor of Leonard Blusse*. Edited by. Nagazumi Yoko. Tokyo: The Toyo Bunko. 253–257.

Cook, James W. (2001). *The Arts of Deception: Playing with Fraud in the Age of Barnum*. Cambridge MA: Harvard University Press.

Frege, Gottlob ([1884] 1953). *The Foundations of Arithmetic*. Translated by J.L. Austin, second revised edition. New York: Harper.

Gaudreault, André (2009). *From Plato to Lumière: Narration and Monstration in Literature and Cinema*. Translated by Timothy Barnard. Toronto: University of Toronto Press.

Kipling, Rudyard (1899). *Stalky & Co*. Project Gutenberg. https://www.gutenberg.org/ebooks/3006

McGuckin, J. (2017). *The path of christianity: The first thousand years*. Downers Grove, IL: InterVarsity Press.

Merleau-Ponty, Maurice (1968). *The Visible and the Invisible*. Translated by Alphonso Lingis. Evanston: Northwestern University Press.

Mulvey, Laura (1975). 'Visual Pleasure and Narrative Cinema'. *Screen* 16(3) Autumn. 6–18.

Origen (1885 [c220]). *De Principiis*. Translated by Frederick Crombie. From *Ante-Nicene Fathers, Vol. 4*. Edited by Alexander Roberts, James Donaldson, and A. Cleveland Coxe. Buffalo, NY: Christian Literature Publishing Co. https://www.newadvent.org/fathers/0412.htm

Quijano, Aníbal (2007). 'Coloniality and Modernity/Rationality', *Cultural Studies*, 21(2–3), 168–178.

Rittvo, Harriet (1997). *The Platypus and the Mermaid*. Cambridge MA: Harvard University Press.

Trimmer, J. D. (1980). The present situation in quantum mechanics: A translation of Schrödinger's 'cat paradox' paper. *Proceedings of the American Philosophical Society*, 124: 323–338.

CHAPTER 15

The Ecological Catastrophe of Algorithmic Individuation: Technological Mediation and Its Social Implication in the Age of the Anthropocene

Jacob Boivin

Introduction

Many have proffered that the ecological transition that is needed today is as much a matter of cultural and psychosocial changes as it is a matter of environmental protection. As early as 1989, Félix Guattari's *Three Ecologies* suggested that the problem we are facing is not simply environmental, but also social and mental. It is more than simply a matter of rationalizing our relation to the environment for a possible sustainable future. The question is one of understanding the production and (re)production of the symbolic representations that constitutes the basis of how humans as 'language animals' (Taylor 2019) entertains different modes of existence,

J. Boivin (✉)
École des Médias, Université du Québec à Montréal, Montréal, QC, Canada
e-mail: boivin.jacob@uqam.ca

D. Binns and R. Najdowski (eds.), *Confronting the Climate Crisis*, Palgrave Studies in Media and Environmental Communication,
https://doi.org/10.1007/978-3-031-89606-4_15

or modes of 'being-in-the-world' to use an Heideggerian term. As much as I agree with this position, I notice that quite often, if the importance of a cultural or psychosocial change is underlined, not much attention is given to the way technology plays a role in articulating the possibilities (or impossibilities) of such a psychosocial change. A certain 'technosolutionism' (Morozov 2014; Baskin 2015) transpires in discourses surrounding the high hopes of new digital technologies and the 'fourth industrial revolution' they bring about (Schwab 2016). The self-proclaimed 'eco-modernists' notably advocate for a 'pragmatic' use of digital technologies which should lead to a decoupling of human activities and their ecological impact. As Leslie Sklair shows (2021), this rhetoric is a dominant discourse throughout the world. In addition to the blind spot of 'rebound effects', this perspective says nothing about the cultural mediation of technology. It does not draw the critical conclusion it should regarding the very technicity of the cultural reproduction. I think such a shift is needed for today's critical ecological thinking. We need to read the technological mediation as an ontological condition. Starting from a phenomenological model inspired by French philosopher Bernard Stiegler, this article posits that technological mediation is always constitutive of our experience of the world, of our 'being-in-the-world'. Therefore, it should be treated, not as '*a*' dimension of the psychosocial conditions, but as '*the*' fundamental one, the one that articulates the very relation between the individual and the collective, and which is therefore at the core of the possible cultural changes that are needed today. To articulate this idea, I start from an understanding of contemporary media inspired by the analysis of media theorist Mark B. Hansen. I apply this understanding and Stiegler's phenomenological model to a reading of what we call the 'algorithmic mediation' that characterizes today's dominant technological mediation. I defend the idea that we are today inscribed in an 'algorithmic individuation', an algorithmic becoming. I then explore how this mode of individuation is exploited by the contemporary structure of digital economy and problematize this as an ecological catastrophe.

Twenty-First Century Media and the Algorithmic Condition

By using the notion of twenty-first century media, we do not claim a perfect homogeneity throughout the domain of media; I simply want to point toward a general condition that can be useful as a theoretical tool.

Mainly, I want to avoid the pitfall of an overly narrow conceptualization of media that would contain our thinking outside of the qualitative difference that nowadays dominant mediation entails. As we will see, twenty-first century media operate based on a digital, and specifically algorithmic, condition that is unique in terms of temporality and signification. I argue that underlining such a difference, regardless of its capacity to operate as a universal concept, can point us toward ways to renew our relation to technology regarding the challenges of the twenty-first century, the condition of the Anthropocene surely being one of them.

In September 2013, a group of physicists published a paper underlining what they called the rise of a new machine ecology functioning beyond the human response time (Johnson et al. 2013: 1). Studying the introduction of high frequency stock trading following 2006, they observed an immense amount of ultrafast extreme events (UEEs, spikes, and crashes) induced by algorithm-to-algorithm interactions. Linking those micro events happening on a scale of milliseconds to the financial collapse of 2008, they argued that these interactions could be understood as an ecological system of machines operating beyond the human time response (which is around one second). Although our interest here is not related to the financial market per se, this example constitutes a good representation of the type of transition we have witnessed and are still witnessing today in terms of technological mediation. What makes these algorithms so efficient is of course their technical ability to process calculation at an extreme speed, but it is also (and probably above all) their fundamental logic and epistemology, being one of anticipation and prediction based on the processing of binary signals into statistical calculation and not on the interpretation of meaningful symbols. What they calculate is the probabilities for a future event to occur through correlational statistics. They then automatically act in anticipation of this said event following a set of preconceived rules, transforming the accurate prediction into earnings in the case of stock trading. This logic is at the very core of today's technological mediation, being social media, high-frequency trading, data mining or, as we will discuss further, the structure of capital accumulation itself. Our daily activities are increasingly mediated through asignifying semiotic mediations (Lazarrato 2014: 40) where the means (the technical abilities of algorithmic analysis) act as finalities.

As media theorist Mark B. Hansen points out, this can qualitatively be understood as a shift from a past-directed recording mediation to a data-driven anticipation of the future (2015: 15). Whereas the mediation of technology used to be coupled with, or even directly synchronized with human sensory apparatus (vision, hearing, etc.), it now operates below that level and even in between technological devices themselves. Therefore, the way this new technological ecology of senses operates must be remediated back to the human time consciousness to be experienced. In a sense, we are always being 'fed forward' what has already been perceived by the algorithmic mediation. This radically questions any remaining instrumental perspective about technology but also opens up to the type of human/machine relations that we can foster regarding where we stand as humans in this new digital and algorithmic ecosystem. Also, and this will be my further argument, this new technological ecology has to be considered as a fundamental element of the question of the broader ecological condition—that of the Anthropocene. Since human activities can be conceived as a disruptive force of geological proportion (Crutzen and Stoermer 2000) affecting the global ecosystems, how then can we interpret the ecological impact of the operation of 'feed-forward' algorithmic mediation?

To attempt to answer this question, I propose to understand the algorithmic condition as a mediation of the structures of thought and subjectivity. Media theorist Matthew Flisfeder echoes this problematic in his book when he says that the 'medium is our metaphor' (2021: 8). He thus inscribes his thinking into the lineage of Marshall McLuhan, who famously stated that 'the medium is the message', which was pointing toward the primary importance of the form of the message against the content (1994: 9). Flisfeder is critical of McLuhan's position and how it tends to reduce meaning to the form of a medium and situates himself closer to Neil Postman's similar affirmation: 'the medium is a metaphor'. But if Postman aims to point out that the dominant medium is a point of inquiry into the dominant discourse of society, Flisfeder goes further and posit that the same can be said for the structures of thought, that is our way of making sense of the world and inscribing ourselves in it through social processes. That is why his book's title (*Algorithmic Desire. Toward a New Structuralist Theory of Social Media*) states the possibility of establishing a new structuralist theory of social media. Being always *in* the media rather than *with* the media (Deuze 2014), living through the mediation of technology, it comes to constitute a condition of our structures of thought. Thinking cannot be fully grasped outside of its technologically mediated condition. This echoes the extended mind thesis of Andy Clark and David Chalmers (1998) as well as the inherent technicity of the posthuman philosophy (Haraway 2007; Sloterdijk 2011; Braidotti 2013; Stiegler 2018). Technology is always ontologically constitutive and influences the capacities of experience as well as of anticipation trough a mediation of the symbolic production of the subject. To offer an understanding of such inherent technicity of subjectivity, I turn toward the post-phenomenology of technology developed by the French philosopher Bernard Stiegler.

A (Post)Phenomenology of Individuation[1]

For Heidegger, being is not understood as a substance (contrary to classical metaphysics), but as a temporal experience. The being is a 'being-in-the-world', a being-there (*Dasein*), and its ontology is therefore linked to its ontic condition as the modalities of an effective realization, a becoming. Gilbert Simondon defines it as an individuation: the perpetuation of the being in time, where the being is but a phase of this process that (re)produces it. The individual emerges as a resolution of pre-individual tensions and thus becomes (the new) pre-individual conditions itself. Time is the essential dimension that constitute the *transductive* relation of the being and its individuation. By transduction, Simondon refers to a relation which constitutes the terms of the relation themselves and articulate their individual reproduction at the same time.

If the *Dasein* must experience itself as a temporal experience, that is as a projection in time (an anticipation), it can only be done from a past that is not its own (not lived) but that is already-there (ontic conditions). The possibilities of access to this already-there, or pre-individual, (its exteriorization and transmission) is what Stiegler theorized as an epiphylogenetic memory. It constitutes a recapitutalive, dynamic, and morphogenetic (phylogenesis) accumulation of individual experiences (epi) (Stiegler 2018: 205). This type of memory requires a form of synthetic objectification of the past that Stiegler identifies as a tertiary retention (technology). With this last concept, he completes Husserl's phenomenological model of primary retention (instant experience) and secondary retention (experience rendered as memory) through a third retention that articulates the relation between the first two and is fundamentally collective since it inscribes the legacy of lived experiences. Tertiary retention is what allows the access to the past necessary for the individuation to take place. Memory is therefore a technical facticity, a prosthetic extension of consciousness that is collective in nature and which Stiegler thus understands as technicity. In this sense, for Stiegler, as for Simondon, individuation is always psychic and collective, but also technological. Individuation is consequently psychic, collective, and technological in a transductive manner. The collective (society, culture) articulates its (re)production through the transmission of memory (technology) and the way it allows individuals to anticipate their becoming and actively realize it.

If culture can therefore be understood as a collective prosthetic extension, there is nonetheless a danger to this inherent technicity, as these and others thinkers have analyzed it. For Heidegger, modern technology presents itself as a pure calculable rationality that transforms the relation between the being and the world such that everything becomes measurable, calculable, and therefore exploitable as a 'standing reserve' (1980). An example is the contemporary shift from possibility to probability (which is echoed in the algorithmic condition). By thinking technology as neutral and simply instrumental, the being is put under the 'dominance of the essence of technology' and is at risk of confounding its own destiny with that of technology. For Simondon, the machines are now 'carrier of tools' (2012: 165) instead of the human, and we are therefore alienated from our own (technically supported) becoming. I hold a similar analysis through Stiegler's reading of technology since the tertiary memory is nowadays industrialized and a rupture is introduced in the triple individuation. The modalities of inscription of memory are less the result of a collective and political dynamic but are instead discharged onto automated modalities specific to technology itself: what I identify today as an algorithmic mediation. Indeed, as I have underlined before, the algorithmic mediation now operates on a pre-individual level that process digital data (tertiary retention) as a way of preempting the becomings through an (a)political and (a)normative rationality (Rouvroy and Berns 2013). This induces an overdetermination of technology upon individuation that I consider to be a major ecological problem since it imposes a form of becoming specifically where we should operate radical ruptures in modes of existence.

Toward Algorithmic Individuation in the Digital Economy

If individuation is always technologically supported, how then can we interpret the algorithmic mediation that I introduced before? Individuation can now be said to be algorithmic in nature (Rouvroy and Berns 2013; Stiegler 2015) since it rests upon an:

> [...] (a)normative or (a)political rationality founded on the automated collection, aggregation and analysis of big data so as to model, anticipate and pre-emptively affect possible behaviours. (Rouvroy and Berns 2013: 173, our translation)

In today's digital economy, this algorithmic mediation is a key technical feature. As underlined by many contemporary theorists (Srnicek 2016; Ouellet 2016; Zuboff 2019), the new dominant structure of capital accumulation rests upon the exploitation of digital data, which is championed by the platforms (Srnicek 2016; Durand 2021). The rise of this new structure of economy followed the massive production of data that accompanied the introduction and evolution of digital technologies to a ubiquitous condition. Becoming quite quickly too much data to be treated by humans, a technological solution imposed itself: data analysis based on an algorithmic mediation (Cardon 2015; Andrejevic 2013). Through the capture of data, its analysis, and its algorithmic processing, it has become possible to operate statistical predictions on consumer's behaviors with an increasing accuracy and to automate the process along the way. These predictions could then be sold to advertising companies and turned the initially not profitable economic model of the tech companies of early 2000 such as Google into highly profitable ventures. Quite quickly, this became the dominant economic model and forced, and still is forcing, a shift in economic structures. The platform turned out to be the ideal structure to nurture a cycle of data production and exploitation through the everyday user. As Nick Srnicek offers, this logic has since influenced other economic sectors such as automobile production, finance, or even agriculture (2016). To a certain point, every daily action is producing digital data and inscribing the most private part of our lives into this new form of capital accumulation.

To understand the relation between the collective individuation and the digital process of capital accumulation, I borrow a schematic analysis of the economic cycle of digital data produced by Ménard et al. (2016) (Fig. 15.1). This will serve as a way of grounding my phenomenological understanding of how the production of data always comes back to the subject in an accelerating cycle of symbolic industrialization that introduces an overdetermination of technological mediation. I follow a certain number of steps in a linear way, but this is strictly an analytical linearity for the written medium. It should indeed be interpreted as a cycle with no specific beginning or ending.

First, the user produces data through any daily actions. Simply being a subject is nowadays associated to producing data which opens the contemporary process of subjectification to datafication (Citton 2016). We can understand the digital data as Stiegler's tertiary retention.[2] It is mnemonic in nature, it is traces of action, which makes it possible to monitor and collect through generalized surveillance (Zuboff 2019). The thing is that this coupling of subjectification and datafication is inscribed in a sociohistorical context,[3] that of an 'hyperindividualism' (Aubert 2006; Mondoux 2011; Stiegler 2013) that rose as the dominant structure of subjectification through what Jean-François Lyotard called the 'end of metanarratives'.[4] Without a pre-individual disciplinary injunction collectively established, the subject is charged with its own self-production. In

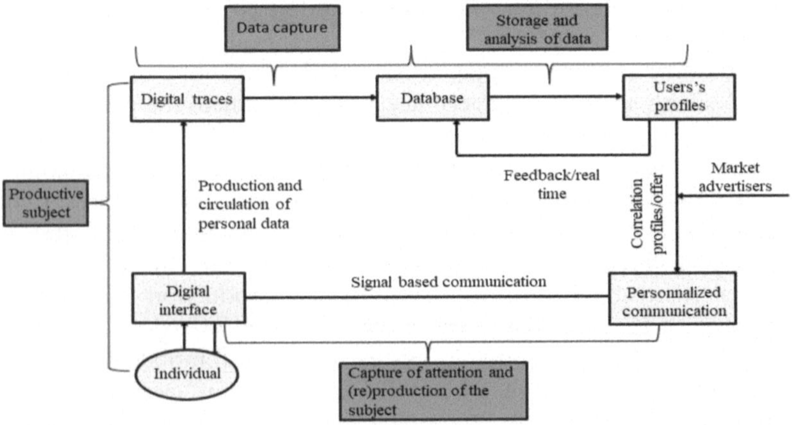

Fig. 15.1 Big data economical cycle applied to digital social media (Ménard et al. 2016)

this context, subjectivity is perceived (and performed) as an act of self-constitution, as a subject giving itself its own normative horizon, freed from the weight of the social (criticized as ideologically normative) and seen as a process of 'empowerment'. In our phenomenological model, this can be understood as the modalities through which the subject encounters the world, that is for and through itself. This participates to shed the concept of society as a totality and to replace it with a technological aggregation of data as a unifying understanding of the social without a finality.

The data that are produced through the actions of the contemporary hyperindividuals are then captured by platforms through different direct or indirect canals. Like we established before, the more the data, better the statistical series, better the statistical prediction and better the control. Therefore, the context of algorithmic mediation also goes along with an always additionally immersive infrastructure of digital sensors to feed this scalability of control. I will address the ecological implications of this condition later, but it suffices to say here that it rests on an always increasing technological infrastructure. Algorithmic mediation's efficacy is based on a form of technicized ontology that constitute a contemporary *universum* (Ellul 2012).

Next come the operation of analyzing and processing these data. This operation is an important mediation since like Lisa Gitelman reminds us 'raw data is an oxymoron' (2013). Any form of data requires a form of labeling, a structural reading of the world and its information. Also, as Srnicek underlines it in his analyses of what he calls a 'platform capitalism', this is the operation that renders the data profitable as a resource since isolated data says nothing about the world, they need to be put in relation with the said world. Being political in nature, this phase is where biases (representations of the world) are most easily reproduced and identifiable in the algorithmic mediation and could notably lead to reproduction and reinforcement of inequalities (Noble 2018; Eubanks 2018). Additionally, it rests on an abstraction of the real that is founded on and adequation between the data and reality which tends to be presented '*as*' reality itself. The data is presented as the real (or considered as such) and not as a reading of it. This im-mediate (without considering the technical mediation itself) interpretation of data based on a dominant neutral interpretation of technology fosters the influence of a technological determinism. Even more, algorithms function on a procedural logic, not a symbolic one (Cardon et al. 2018). The symbolic dimension of the

The Ecological Catastrophe of Algorithmic Individuation

The ecological impact of the algorithmic condition of individuation that I just described can be circumscribed into three different aspects that are linked together: a material (geochemical), a biological (biodiversity) and a psychosocial destruction. On the first level, we must tackle what Jussi Parikka identifies as the 'geology of media' (2015). We need to understand, contrary to the ideology of the immateriality of the digital, that media technologies are always intertwined with natural processes. In that sense, the very possibilities of the existence of a media environment fully constituted through and by an algorithmic logic is already questionable. The need for an always growing amount of data following the importance of scalability in the digital economy is linked to an always growing amount of processors and infrastructures that require an important quantity of resources and energy. Indeed, the evolution of the power of processors that made algorithms a digital technology so central today is followed by an equivalent increase in the consumption of rare metals and natural resources. The Material input per service unit (MIPS) which is used to measure the quantity of material needed to produce a good or a service is incredibly high for today's microchips. Thirty-two kilograms of matter is needed for an integrated circuit of two grams, a ratio of 16,000/1 (Bordage 2019). Also, the number of raw materials needed to produce a phone has grown between 1960 and 2021 from 10 elements for landline phones to 57 elements for smartphones (Pitron 2021: 330). This includes metals such as cobalt, lithium, germanium, dysprosium and neodymium and the extraction of theses rare metals represents only about 0.2% of the amount of earth that needs to be mined to acquire them (Crawford 2021: 37). These resources are not renewable and are most of the time extracted in terrible conditions either in terms of human labor or impact on biodiversity. In the case of cobalt for example, the blasting methods and the energy needed for the extractions are an important source of ecological disturbances (Farjana et al. 2019). The surge in global mining brought about by the hopes of general smartification thus imposes increasing pressure on ecosystems and erodes biodiversity (Luckneder et al. 2021: 2) while digital technologies fail to reduce energy demand (Lange et al. 2020: 8). In addition, an important amount of energy is needed to 'train' the needed artificial intelligence models (AI). OpenAI estimates that since

2012, the amount of computing needed to train large AI models has been increasing exponentially, doubling every three to four months (OpenAI May 2018). Estimating the cost of this computing power in terms of carbon dioxide emissions, Emma Strubell and her colleagues concluded that training a version of Google's language model, BERT,[6] produced 1438 pounds of CO_2 (Strubell et al. 2019). This should also be linked to the increase in data centers that are now among the world's largest consumers of electricity (Luccioni et al. 2023).

In addition to this material (geological) and biological (biodiversity) destruction, the ecology of the algorithmic condition needs to be thought through the psychosocial influence that I underlined. As we have seen, it operates a circular 'self-inscription' into the cycles of capital accumulation through the datafication of subjectivity. But the ecological danger is not simply one of industrializing symbolic representations and attention capacities as could be discussed through and 'ecology of attention' (Citton 2014). It is also, and especially, the way this is linked to the material and biological cost of the algorithmization of the world. By operating on an apolitical and anormative level that bypasses the collective individuation, the algorithmic mediation is limited to a logic of optimization by calculation that tends to reproduce the status quo or even feed a certain 'path dependance' (Green 2019). The problematic that I identified through the acceleration of the digital data cycle leads to an increasing algorithmic mediation, which leads in return to a spike in resources exploitations, data centers constructions, digital infrastructure, biodiversity destruction, etc. The COVID-19 pandemic, which can be linked to the destruction of biodiversity and forests (Lawler et al. 2021), has incidentally accelerated such a cultural influence of digital technologies. Naomi Klein soundly criticized the 'Screen New Deal' that has unfolded instead of the expected, and needed, 'Green New Deal' (May 13 2020). Indeed, the hegemony of platforms has been solidified through the last years of crisis and they seem to have emerged as the only 'winners',[7] without making a dent into the ecological impact of human activities. This ongoing (since 2008 at least) restructuration of socioeconomic structures around platforms and their use of algorithmic mediation should therefore be inscribed in the horizon of an ecological catastrophe since it only deepens the inscription of subjects through capital accumulation and ecological destruction. By bypassing the very condition of the political (collective individuation), it becomes increasingly difficult to operate the radical cultural changes needed. Therefore, the material (geochemical), the biological (biodiversity) and the psychosocial destruction at the basis of the Anthropocene are linked together in a downward spiraling process.

Reopening Potentialities, the Psychosocial as a Prosthetic for the Technological

In addition to a material insolvency, the algorithmic mediation upon which rests the logic of big data can also be understood as a psychosocial insolvency since the accelerating coupling of technology that I analyzed induces a destruction of the individuation process that constitutes the basis for a radical change in modes of existences. In this sense, individuation can be understood as a form of 'resource' of sort, that is being exploited and destroyed just as natural resources. Nevertheless, it cannot be exhausted since it is not material. Individuation can be understood, following Gilbert Simondon, as a fundamentally metastable condition. It is the perpetual resolution of the tensions that constitute our experience of time and is therefore where we should anchor the possibilities of an ecological transition. If we don't recognize the technically supported condition of our becoming (prosthetic to use Stiegler's term), we cannot fully apprehend the ecological implications of a socioeconomic model based on an anticipating technological mediation. Our processes of individuation are being 'bypassed' by the foreclosing of a predicted future upon the possibilities of the present. As the sociologist Dominique Cardon says regarding the logic of algorithms, 'the probable preempts the possible' (2015: 89). However, in the context of the Anthropocene, the necessity is to break free from a 'business as usual' relation to the future, it is to open our ways of living, which are always technologically supported becoming, to new possibilities. That is why I bring the attention of ecological critical thinking toward the importance of technology as a mediation of the structures of thought and experience as a necessary way of grasping contemporary issues regarding technological development.

As I said earlier, individuation is founded upon a pre-individual condition that is an open potentiality. In this sense, subjectivity can never be fully reduced to its technological rendering (data). It is always something more, and that should be interpreted as a foundation for an ecological and political project based on a critical understanding of predictive epistemology. As the prosthetic condition of contemporary individuation, the algorithmic mediation tends to foreclose this potential and should therefore be criticized on this ground. One way seems to me to conceive of the individuation potentialities as a form of 'prosthesis for the technical prosthesis'. The technological is, in a sense, psychosocially lacking just as the human being is technologically lacking. Where technology constitutes a support to the individuation process, it is also a danger to individuation if not redoubled by a 'psychosocial prosthesis', a human critical prosthesis for the limits of the technological condition. Since it operates on a (a)normative and (a)political correlational epistemology, it is lacking a symbolic component inherent to human modes of cognition. It is that component that serves as the foundation of critical thinking since it cannot do without normative decisions. This means to place technologies at the center of the political, and especially at the heart of a psychosociopolitical ecology. In turn, this also means operating a critique of datafication as subjectification and how this coupling tends to industrialize symbolic representations, which are the basis for the constitution of a political project. The latter should therefore constitute of multiple social and local terrains aimed at experimenting and fostering individual and collective new relations regarding twenty-first-century media which would channel the potentiality of individuation instead of foreclosing it. Only starting from this rebuilt relation to technology will we be able to address its material, social, and psychological impact in a truly ecological sense.

Notes

1. The (post) here is in parenthesis because Stiegler himself never referred to his model as being postphenomenological since his whole understanding is founded on the idea that a phenomenological account is inherently technological. Here, his understanding differs from the contemporary current that explicitly positions itself as postphenomenological and is championed by figures such as Don Idhe, Robert Joseph Rosenberg or Peter-Paul Verbeek. They offer more of a turn toward technology as a starting point, a 'philosophy from technology', rather than a rethinking of phenomenological experience through an inherent technicity.
2. With the advent of fully automated and algorithmic technological systems, it is however possible to speculate on the relevance of a conceptual 'quaternary retention' that would designate an

exclusively technological logic of memory inscription which would operate a rupture into the phenomenological reading of technicity. A certain critique of the Stiglerian model could then be traced along those lines and is indeed echoed in Hansen's critique regarding the fact that Stiegler retains human consciousness as the exclusive site for thinking technology (2015: 78) even though he addresses the temporal 'short circuiting' of digital technologies regarding the time of consciousness.

3. In echo to our posture of individuation, it is indeed impossible to try to understand a subject outside of its timely inscription. This is sometimes neglected in analyses that concentrates on the usage of technology.
4. Meta-narratives are understood as sorts of symbolic totalities, such as communism, the notion of progress or the Enlightment, which would inscribe individuals into a definite collective horizon. Post-modernity is then understood as a failure of these narratives to deliver on their promises. It leads to what Slavoj Zizek identified as the 'demise of the symbolic' and what Bernard Stiegler conceptualized as a 'symbolic misery'.
5. Since totality cannot be grasped as an object (seen from the 'outside'), it is in the realm of representations and ideology. Thus, the end of totality is also the end of 'meta-narratives' and the occultation of the political, clearing the path for operational process (capital's accumulation dynamic) and the reliance on a seemingly neutral technology.
6. Which stands for Bidirectional Encoder Representations from Transformers.
7. Just like how social media's structural influence seems to have been the biggest impact of what has since been called 'social media revolutions'.

Bibliography

Andrejevic, Mark (2013), *Infoglut: How too Much Information is Changing the Way we Think and Know*, London: Routledge.

Aubert, Nicole (2006), *L'individu hypermoderne*, Toulouse: Erès.

Baskin, Jeremy (2015), 'Paradigm Dressed as an Epoch: The Ideology of the Anthropocene', *Environmental Values*, 24:1, pp. 9–29.

Bordage, Frédéric (2019), *Empreinte environnementale du numérique mondial*, GreenIT.fr, https://www.greenit.fr/wp-content/uploads/2019/10/2019-10-GREENIT-etude_EENM-rapport-accessible.VF_.pdf, Accessed 10 January 2022.

Braidotti, Rosi (2013), *The Posthuman*, Cambridge: Polity.

Cardon, Dominique (2015), *À quoi rêvent les algorithmes. Nos vies à l'heure des big data*, Paris: Seuil.

Cardon, Dominique, Cointet, Jean-Philippe et Mazières Antoine (2018), 'La revanche des neurones. L'invention des machines inductives et la controverse de l'intelligence artificielle', *Réseaux*, 5:211, pp.173–220.

Citton, Yves (2014), *Pour une écologie de l'attention*, Paris: Seuil.

Citton, Yves (2016), 'Subjectivités computationnelles à l'erre numérique', *Multitudes*, 1:62, pp.45–64.

Clark, Andy and Chalmers, David (1998), 'The extended mind', *Analysis*, 58: 1, pp.7–19.

Crawford, Kate (2021), *Atlas of AI*, New Haven and London: Yale University Press.

Crutzen, Paul, Stoermer, Eugene (2000) 'The Antropocene', *Global Change Newsletter*, 41, pp.17–18.

Deuze, Mark (2014), *Media Life*, Hoboken: Wiley.

Durand, Cédric (2021), *Techno-féodalisme. Critique de l'économie numérique*, Paris: La Découverte.

Ellul, Jacques (2012), *Le système technicien*, Paris: Cherche-midi.

Eubanks, Virgina (2018), *Automating Inequality: How High-Tech Tools Profiles, Police, and Punish the Poor*, New York: Macmillan.

Farjana, Shahjadi, Hisan, Huda, Nazmul and Mahmud, Parvez (2019), 'Life Cycle Assesment of Cobalt Extraction Process', *Journal of Sustainable Mining*, 18:3, pp.150–161.

Flisfeder, Matthew (2021), *Algorithmic Desire. Toward a New Structuralist Theory of Social Media*, Evanston: Northwestern University Press.

Gitelman, Lisa (2013), *Raw Data is an Oxymoron*, Cambridge: MIT Press.

Green, Ben (2019), *The Smart Enough City. Putting Technology in its Place to Reclaim Our Urban Future, Cambdrige*: MIT Press.

Hansen, B. Mark (2015), *Feed-Forward. On the Future of Twenty-First Century Media*, Chicago and London: The University of Chicago Press.

Haraway, Donna (2007), *Manifeste cyborg et autres essais: sciences, fictions et féminismes*, Paris: Exils.

Heidegger, Martin (1980), *Essais et conférences*, Paris: Gallimard.

Johnson, Neil, Zao, Guannan, Zhao, Hunsader, Eric, Qi, Hong, Johnson, Nicholas, Meng, Jing and Tivnan, Brian (2013), 'Abrupt Rise of New Machine Ecology Beyond Human Response Time', *Scientific Reports*, 3: 2627, pp.1–7.

Klein, Naomi (2020), 'How Big Tech Plans to Profit from the Pandemic', *The Guardian*, 13 May, https://www.theguardian.com/news/2020/may/13/naomi-klein-how-big-tech-plans-to-profit-from-coronavirus-pandemic, Accessed 12 January 2022.

Lafontaine, Céline (2004), *L'empire cybernétique. Des machines à penser à la pensée machine*, Paris: Seuil.

Lange, Steffen, Pohl, Johanna and Santarius, Tillman (2020), 'Digitalization and Energy Consumption. Does ICT Reduce Energy Demand?', *Ecological Economics*, 176, pp.1–14.

Lawler, Odette K, Hannah, L Allan, Peter, W J Baxter, Romi, Castagnino, Marina, Corella Tor, Leah, E Dann, Joshua, Hungerford, Dibesh, Karmacharya, Thomas, J Lloyd, María, José, López-Jara, Gloeta, N Massie, Junior, Novera, Andrew, M Rogers, Salit, Kark (2021), The COVID-19 Pandemic is Intrincately Linked to Biodiversity Loss and Ecosystem Health, *Lancet Planet Health*, 5: pp. e840–850.

Lazarrato, Maurizio (2014), *Signs and Machines. Capitalism and the Production of Subjectivity*, Cambridge: MIT Press.

Luccioni, Sasha, Jernite, Yacine and Strubell, Emma (2023), 'Power Hungry Processing: Watts Driving the Cost of AI Deployment', *arXiv*, arXiv:2311.16863. Accessed 20 May 2025.

Luckneder, Sebastian, Giljum, Stefan, Schaffartzik, Anke, Maus, Victor and Tost, Michael (2021), Surge in Global Metal Mining Threatens Vulnerable Ecosystems, *Global Environmental Change*, 69, July, pp.1–14.

Massumi, Brian (2015), *Ontopower: War, Powers, and the State of Perception*, Durham: Duke University Press.

McLuhan, M. (1994), *Understanding Media: The Extension of Man*, Cambridge: MIT Press.

Ménard, Marc, Mondoux, André, Ouellet, Maxime et Bonenfant, Maude, (2016), *Big Data*, gouvernementalité et industrialisation des médiations symboliques et politico-institutionnelles, *Interfaces Numériques*, 5: 2.

Mondoux, André (2011), *Histoire sociale des technologies numériques de 1945 à aujourd'hui*, Montréal: Nota Bene.

Morozov, Evgeny (2014), *To Solve Everything, Click Here. The Folly of Technological Solutionism*, New York: Public Affairs.

Noble, Safya (2018), *Algorithms of Oppression. How Search Engines Reinforce Racism*, New York: NYU Press.

OpenAI (2018), 'AI and Compute', *OpenAI*, 16 May, https://openai.com/blog/ai-and-compute/, Accessed 20 January 2022.

Ouellet, Maxime (2016), *La révolution culturelle du capital. Le capitalisme cybernétique dans la société globale de l'information*, Montréal: Écosociété.

Parikka, Jussi (2015), *A Geology of Media*, Minneapolis: University of Minnesota Press.

Pitron, Guillaume (2021), *L'enfer numérique: Voyage au bout d'un like*, Paris: Les liens qui libèrent.

Rouvroy, Antoinette et Berns, Thomas (2013), 'Gouvernementalité algorithmique et perspectives d'émancipation. Le disparate comme condition d'individuation par la relation ?', *Réseaux*, 1: 177, pp.163–196.

Schwab, Klaus (2016), *La quatrième révolution industrielle*, Malakoff: Dunod.

Simondon, Gilbert (2012), *Du mode d'existence des objets techniques*, Paris: Aubier.

Sklair, Leslie (2021), *The Anthropocene in Global Media. Neutralizing the Risk*, London: Routledge.

Sloterdijk, Peter (2011), *Tu dois changer ta vie!*, Paris: Libella Maren Sell.

Srnicek, Nick (2016), *Platform Capitalism*, Cambridge: Polity.

Stiegler, Bernard (2013), *De la misère symbolique*, Paris: Flammarion.

Stiegler, Bernard (2015), *La société automatique 1. L'avenir du travail*, Paris: Fayard.

Stiegler, Bernard (2018), *La technique et le temps. 1. La faute d'Épiméthée – 2. La désorientation – 3. Le temps du cinéma et la question du mal-être suivis de Le nouveau conflit des facultés et des fonctions dans l'Anthropocène*, Paris: Fayard.

Strubell, Emma, Ganesh, Ananya and McCallum, Andrew (2019), 'Energy and Policy Considerations for Deep Learning in NLP', *Proceedings of the 57th Annual Meeting of the Association for Computational Linguistics*, Florence, Italy, July 28–August 2, pp. 3645–3650.

Taylor, Charles (2019), *L'animal langage. La compétence linguistique humaine*, Montréal: Boréal.

Vioulac, Jean (2018), *Approche de la criticité: philosophie, capitalisme, technologie*, Paris: PUF.

Zuboff, Shoshana (2019), *The Age of Surveillance Capitalism. The Fight for a Human Future at the New Frontier of Power*. New York: Public Affairs.

Index

D. Binns and R. Najdowski (eds.), *Confronting the Climate Crisis*, Palgrave Studies in Media and Environmental Communication, https://doi.org/10.1007/978-3-031-89606-4

The manufacturer's authorised representative in the EU is Springer Nature Customer Service Centre GmbH, Europaplatz 3, 69115 Heidelberg, Germany. If you have any concerns regarding our products, please contact ProductSafety@springernature.com

Printed and bound by CPI Group (UK) Ltd, Croydon, CR0 4YY
30/09/2026
02225567-0002